Easy Solution of Practical Transmission Line Problems

A Calculation Manual for Amateur Radio Operators

William Troy Cowan

ACØFV

Easy Solution of Practical Transmission Line Problems
A Calculation Manual for Amateur Radio Operators

Printed in the United States of America
ISBN: 147926363X
ISBN 13: 9781479263639

ABOUT THE AUTHOR

William Cowan was first licensed as an amateur in 1954 and Licensed as a first-class radio telephone operator in 1959. He earned the BS and MS in electrical engineering in 1961 and 1965, respectively.

In February 1965, he joined the Hewlett-Packard Company research and development laboratory in Loveland Colorado as an electronic instrument design engineer. During a career which spanned more than thirty years with HP he held many positions including design engineer, project manager, production section manager, and division functional manager, finally ending as corporate manufacturing engineering manager located at the company headquarters.

Will has written and published two short academic books concerning specific antennas, several company internal technical publications, and an article that appeared in *HP Journal*.

PREFACE

Learning about electricity and electronics virtually always begins with the study of Ohm's Law applied to direct current circuits. Most often this is followed by the study of reactances, impedances, and then resonance in alternating current circuits. Those who advance through these steps often have the belief that Ohm's Law and the simple power equations $P = I * E$ and $P = I^2 * R$ will enable them to calculate all of the voltages, currents, and power levels for virtually every circuit they might encounter. After all, these relationships have always worked for them before for complex circuits involving loop equations, as well as for the simple ones.

Imagine then, their surprise when they first try these techniques upon a circuit which involves transmission lines!

The typical reaction to this experience is to say, "I must have done something wrong." Rarely does the student realize when transmission lines are added to the circuit, that the desired solutions require new and more powerful tools and that simple equations and algebra alone are no longer sufficient to find the desired answers. Several complex mathematical functions as well as the careful use of algebra are necessary to calculate the desired unknowns.

This book introduces the reader to that world of transmission line circuits where a more advanced approach is required and shows how those desirable but elusive quantities can be found after all.

Welcome to this new part of the study of electronics.

Will Cowan
Longmont, Colorado

TABLE OF CONTENTS

1 INTRODUCTION

Those seeking insight into circuits which employ direct current or low frequency alternating current are generally able to ignore the effects of the wires which connect the circuit's source with its load. When the length of these interconnecting wires begins to approach the wavelength of the source signal, such an assumption is no longer valid. Where small resistive losses usually characterize the first case, long interconnecting wires can be inductive and or capacitive and possess high impedances which vary with frequency and interact with the impedances of other connected elements. Because long interconnecting wires possess behaviors different from short inter-connecting wires they are called transmission lines.

Though transmission line behavior is considerably more complex than that of short interconnection wires, transmission line behaviors can be described mathematically and their exact solutions can be determined. Unfortunately, complete understanding of these exact solutions requires some knowledge of advanced mathematics. Consequently, for decades the sources of knowledge concerning transmission lines, for those lacking the necessary mathematical back-ground, have been word-of-mouth conversations, third party experiences, unsubstantiated experiments and wordy descriptive articles that attempt to explain them while avoiding the use of any mathematics. Is it any surprise then, that transmission line users sometimes apply unjustified approximations, purchase unneeded equipment and on occasion have a just plain wrong understanding of them?

This book faces the mathematics issue straight on. It employs the exact mathematical solutions, not approximations; and because obtaining numerical results using the exact solutions is long and involved, computer programs are employed to make these calculations directly. (These programs are included on a CD called *Easy Solution of Practical Transmission*

1

Line Problems, The Disk.[1] This CD contains the programs and the examples described throughout this book, its purchase is strongly recommended.) The programs include easy-to-use data entry forms to populate the exact formulas used by the programs. Because the computer performs all of the calculations, one need not fully understand the program or the mathematical functions they employ to obtain correct results. However, this book includes all of the equations used by the computer, allowing those who are interested to know what the programs are doing and to verify results as they may desire.

Historical Perspective

Telegraphic communications having been established in America and in Europe by the early nineteenth century, the idea of extending such communication between the two continents was born about 1830. Several attempts were made to place cables under the sea to bring this about. However, these efforts met with numerous difficulties. The first problem was the lack of good conductors and good insulation materials, but over the succeeding two decades noticeable improvements were made in these areas. Significant problems remained however. High-voltage sources were required to power the signal over such a long distance; and telegraphic key operators had difficulty dealing with large voltage spikes, signal delays, echoes and garbled signals. Long-distance telegraphic communication remained unreliable. The basic problem was that the physics of the long undersea cable was just not understood. Though the concepts of resistance, capacitance, inductance and several fundamental electrical theorems had been developed by this time, no one had brought sufficient science to bear upon the long cable problem.

In 1873, the brilliant James Clerk Maxwell published one of his most famous works. In his treatise, Maxwell was able to formulate four vector differential equations and three supporting relationships which describe virtually all electrical phenomena in every media. His work also brought together for the first time Coulomb's, Faraday's, Kirchhoff's, and Ohm's laws. His equations covered electrostatics and electromagnetic wave propagation in single or multidimensional space as a function of time and in conducting or lossless environments. His equations remain today as the basis of everything electrical and magnetic in our world. Indeed researchers in the twenty-first century still apply boundary conditions

---

[1] Developed, coded and produced by the author in 2012 and available through Amazon.com.

to solutions of Maxwell's equations to solve modern electromagnetic problems.

Maxwell's equations provided solutions for voltage and current. These solutions included electrical energy, power, time, phase velocity, and factors pertaining to the physics of materials and the environment being used; but they did not specifically include electrical impedance or admittance, because these concepts were yet to be postulated.

In 1886, using Maxwell's general equations, Oliver Heaviside, developed a specific differential equation which included the long undersea cable problem. As with all differential equation solutions, boundary conditions (length, configuration, dimensions, conductor materials, insulation materials, sea water conductivity, time etc.) had to be applied to the solution of his equation in order to make it fit the undersea cable situation. When he completed the solution and applied these undersea boundary conditions, his work became known as the Telegrapher's Equation Solution.

As an aside to his work, Heaviside realized that some of the boundary condition parameters could be combined in ways such that a new quantity that he called *impedance*, could be defined. Over time and after rigorous analysis, this impedance became one of the three equally important electrical quantities that, today are taken for granted: voltage, current *and* impedance.

With the passage of time and the changing use of such cables, Heaviside's original equation became known as the Transmission Line Equation and remains so named today. It has since been solved many times and in a number of ways. Today there are two major approaches to its solution: 1) electromagnetic wave equation solution with appropriate application of boundary conditions and 2) using many small differential equivalent circuit network cells integrated over the line's length thus reaching a mathematical limit. Because the first method is considerably more difficult and because Heaviside's invention of impedance has provided such a helpful tool, the second method is now used most often. Its solution appears in various forms in the work of authors such as David Cheng, Walter Johnson, Ronald King and John Kraus. Because the purpose of this book is to use the equation solution to explore the behavior of transmission lines, not to develop such equations, no derivation of a solution will be attempted here.

Conceptually, applying boundary conditions to the equation solution is straight-forward, but it can appear daunting to the first timer because the solutions are combinations of exponential functions and hyperbolic-trigonometric functions, or their various forms, all with complex arguments and complex coefficients. There is no mystery in bringing these solutions to numerical values for real transmission line applications, but before electronic computing devices, doing so was laborious, time consuming and prone to cumulative calculation errors. These difficulties have hampered the understanding of the physics of transmission line behavior and of their parameters. Consequently, those with a limited understanding of the mathematics of transmission line theory sometimes still think of the Transmission Line Equation solution as mysterious and strange.

The Smith Chart

During the 1920s and 1930s, Phillip H. Smith developed a clever graphical calculator that was invaluable in determining the answers to a number of transmission line problems. His calculator came to be known as the Smith Chart. It afforded an improved understanding of the behavior of transmission lines to the uninitiated than had existed before and provided numerical answers to many problems. Users could also study the chart and gain insight into use scenarios. However, the precision and accuracy of answers obtained via the Smith Chart are often unsatisfactory.

Excel Spreadsheet

The lack of good computational tools was a serious impediment to producing numerical results, and it sometimes hampered understanding of the transmission line's behavior because the numerical results were so difficult to obtain. Personal computers now provide the tool to make the calculations. The question remaining is which software should be used. Today, Microsoft Excel is generally available and possesses scores of computation routines, powerful complex arithmetic functions, and the Visual Basic for Applications language (VBA) making the choice easy. Consequently, all of the programs described in this book have been designed to operate with Excel 2003 and later versions.

The solution to the Transmission Line Equation and its supporting formulas have been programmed into six separate Excel applications specifically for this book, each aimed at separate areas of transmission line usage. Use of each application is simple:

1) Open the blank template file for the particular area of interest.
2) Click the button on the display screen.

3) Enter the input data into the form which results from 2).
4) Click the **Calculate** button shown on the display.
5) Print or save the results.

Details for each application are included later but all of them follow these five steps.

Purpose

The purpose of this book is to present easy-to-use methods that calculate mathematically correct numerical values for various transmission line problems, use these methods to solve a series of practical problems, and demonstrate actual transmission line behavior by discussing the results and implications of the solutions obtained.

Though the book includes the full Transmission Line Equation solution and its supporting relationships, the reader can easily employ the methods contained here without being concerned with the complications of the mathematics of the equation by using the computer applications. However, because the equations are described in some detail, readers may become as involved with the mathematics as they desire.

⌘   ⌘   ⌘

2 DEFINITIONS AND CONDITIONS

Because the mathematical descriptions of transmission lines are complicated, the uninitiated sometimes think that they are mysterious and strange. The fact is that many electrical phenomena are occurring within an operating transmission line. Transmission lines carry AC signals; they possess both capacitive and inductive reactances; these reactances vary with frequency and length of the line; voltage and current magnitudes and phase angles vary along the line's length. When mismatched to the source and or the load, stationary waves exist upon the line and there will be reflections at the line ends causing impaired energy transfer; and in general, signals become weaker as the transmission line length increases. Whew! If one intends to understand transfer of energy via transmission lines one must understand these issues. By studying the text and examples in these chapters, transmission lines should become less mysterious.

What Is a Transmission Line?

The Introduction gave us a start for defining a transmission line. To qualify as a transmission line, the wires connecting a source and its load must be of a length which is not negligible when compared with the source signal's wavelength. However, to fit within our definition more is required. The transmission lines we will discuss transport only alternating current energy from one place to another. This energy is limited on the low end to frequencies in the audio range and has a practical upper limit of the very high frequencies in the Giga-Hertz range. In addition our transmission lines will consist of only two conductors, and these can be of only two types: coaxial cables and balanced lines (i.e., two conductors equally sized and separated by a fixed uniform distance and surrounded by a uniform medium).

Calculation Accuracy

The main contribution of this book is that it provides numerical answers concerning actual transmission lines in use, not just a basis for supporting intuitive feelings about them. By obtaining these actual circuit quantity values and numbers, a basic understanding of transmission line behaviors can be made more clear. In other words, this book is about understanding through calculation. Therefore, it seems appropriate to say a few words about the available methods of calculation that might be used to obtain the desired results.

A typical engineering slide rule usually provides solutions to vectors, exponential, trigonometric and hyperbolic trigonometric functions directly, but its accuracy and precision is only about 3 decimal digits at best. This means that by performing the required series of calculations necessary to arrive at typical transmission line quantities, the final results will not likely possess even 2 significant digits of accuracy. For this reason seeking solutions via slide rules is not recommended.

Calculators can produce about 10 digits of display with a usual accuracy of about 11 or 12 digits, but not all of them can provide the required functional values. This means that tables of exponential, trigonometric, and hyperbolic trigonometric functions must be used. Unfortunately, when such tables are used, the required table extrapolation often diminishes the precision considerably, perhaps resulting in no better than four significant digits of precision. In addition, the use of tables can be prone to error, plus require significant time to reach a solution. Hence, when solving transmission line problems via calculator, the calculator needs to be of an advanced or engineering type which calculates functional values without the need for tables or extrapolation.

Excel provides all of the required functional values, requires no extrapolation, and normally computes each quantity using 15 decimal digits of precision. In some cases in this book example results will be displayed using up to 15 digits. When writing this book, it was decided not to arbitrarily terminate these long computer generated results, because it was reasoned that some readers might want to see the effects caused by small changes in one or more input variables.

It is, however, important to remind readers that no algorithm produces answers that are more *accurate* than the accuracy of any one or all of the constants and variable data used as inputs. If data are known to an accuracy of only 4 significant digits, the results cannot be more *accurate* than 4 significant digits no matter how many digits are used to calculate the results or how many are displayed. Therefore, before claiming levels of calculation accuracy, first ascertain the precision and accuracy of each quantity that is used as input data for the calculation.

How to Use This Book

This work provides the reader with an understanding of how transmission lines behave when they are connected between a source and a load such as a transmitter and an antenna. The principle method used to accomplish this is to calculate circuit quantities such as line and load voltages, currents, impedances, standing wave ratios, reflection coefficients, antenna power, transmitter power, line power loss, etc. The book demonstrates how to calculate these quantities and discusses and interprets the results so obtained. To do all of this necessarily implies that numerous formulas have been preprogrammed into the easy-to-use applications; but should a reader desire to perform the calculations manually, these equations are provided in the text.

The book begins with an overview of transmission line quantities and circuit definitions. Descriptions of the computer applications and how to use them are then presented. The next two chapters cover the areas of everyday problems faced by radio amateurs and include numerous examples. In many cases, the reader's specific problems can be solved by directly entering the data from their own situation into the input form of one of these examples and clicking the **Calculate** button on the data input form. Consequently, it is strongly recommended that the reader study all of the examples in detail.

Should readers discover that they lack one or more pieces of transmission line information, then chapters 7, 8 and 9 will help them determine the needed line parameters.

Chapters 10, 11 and 12 cover special cases of transmission line usage. They include the use of cable fragments as impedance components, and as devices which can match transmission line impedances to antenna impedances.

Amateurs frequently run into the need to convert a quantity expressed in one unit into a quantity expressed in a second unit (e.g., to convert feet into meters or centimeters), but users of transmission lines soon discover that unit conversion is even more important than for other areas of study. Appendix A Conversion of Units of Measure includes several conversion factors, but most important, it includes a tutorial on the process of units conversion and how to derive needed conversion factors.

Because many antennas can be and are matched to their feeder trans-mission lines by using an L-C matching network, Appendix B shows the derivation of equations, which calculate the necessary values of L and C for this purpose

⌘ ⌘ ⌘

3 APPROACHING THE TRANSMISSION LINE EQUATION

All conductors possess series resistance and inductance; and when two conductors are in proximity to one another, shunting capacitance and conductance also always exist between them. Each of these four parameters exist no matter the length of the two conductors; thus we can think of them as possessing distributed resistance, capacitance, inductance and conductance and deal with them on a per unit length basis. As in the lumped case, the magnitudes of these four quantities are determined by the types of materials making up the transmission line, the physical sizes and spacing of conductors and the environment in which they operate.

Before calculations can be made, further quantification is required; we need the actual material conductivities, insulation characteristics and specific conductor conductivity, sizes and spacings, etc. When these quantities are known the **Transmission Line Equation** solution introduced in Chapter 1 becomes a useful equation. This chapter describes these various quantities and shows how they fit within transmission line systems.

The series impedance Z per unit length of transmission line is:

$$Z = R + j\omega L$$

and the shunt admittance Y per unit length of transmission line is:

$$Y = G + j\omega C$$

Where:

R = series resistance in ohms/meter;
L = series inductance in henries/meter;

ω = 2π f = angular velocity in radians/second;
f = the signal source frequency in Hz;
G = shunt conductance in siemens/meter and
C = shunt capacitance in farads/meter

When Z and Y are combined, a key transmission line quantity is defined. It is called the line's characteristic impedance, which uses the symbol Z_o.

$$Z_o = \sqrt{\frac{Z}{Y}} = \sqrt{\frac{R + j\omega L}{G + j\omega C}}$$

For most commercial transmission lines Z_o is a mathematically real quantity, achieved because there is either a good balance of parametric quantities in the numerator and denominator under the radical or because the reactive quantities are much larger than the real quantities.

A second important quantity defining the behavior of a transmission line is its propagation constant γ.

$$\gamma = \sqrt{ZY} = \sqrt{(R + j\omega L)(G + j\omega C)}$$

The propagation constant is most often complex and can be separated into its two components, a real part and an imaginary part. The real part is

$$Re[\gamma] = Re\left[\sqrt{ZY}\right] = \alpha$$

α is called the transmission line attenuation and is a measure of how the signal or wave strength declines in magnitude as it moves along the line. Attenuation is expressed in nepers per meter. (Nepers are described in Appendix A.) The imaginary part is

$$Im[\gamma] = Im\left[\sqrt{ZY}\right] = \beta$$

β is the phase shift experienced by the signal as it moves along the transmission line and is expressed in radians per meter.

A third important quantity is the velocity v with which the signal or wave travels along the line. It is

$$v = \frac{\omega}{\mathrm{Im}[\gamma]} = \frac{\omega}{\beta} \quad \text{in meters} \,/\, \mathrm{sec}\, ond \qquad \{3-1\}$$

Because α and β are both functions of the materials used in the construction of the transmission line, the velocity is also a function of the transmission line materials. Thus those factors relating the line materials to velocity are of interest. From Kraus (Kraus 1953: 352):

$$v = \frac{1}{\sqrt{\mu\varepsilon}}$$

where:

μ = permeability of the medium in Henrys/meter
ε = permittivity of the medium in Farads/meter.

In general, both of these parameters consist of an absolute value and a relative value, i.e.,

$\mu = \mu_o \mu_r$ and
$\varepsilon = \varepsilon_o \varepsilon_r$

In free space both relative terms are equal to 1, i.e.: $\mu_r = 1$ and $\varepsilon_r = 1$, which means $\mu = \mu_o$ and $\varepsilon = \varepsilon_o$. Thus the velocity of any electromagnetic wave in free space can be computed:

$$v = \frac{1}{\sqrt{\mu_o \varepsilon_o}}$$

$$= \frac{1}{\sqrt{\left(4\pi \ x \ 10^{-7}\right)\left(8.854187 \ x \ 10^{-12}\right)}}$$

$$= 2.99792 \ x \ 10^8 \ meters \,/\, \mathrm{sec}$$

$$= c$$

where c is the velocity of light in free space that we are all familiar with. Clearly then, the velocity of a wave or signal within a transmission line will always be less than it is in free space by the amount:

$$v = \frac{c}{\sqrt{\mu_r \varepsilon_r}} \qquad \{3-2\}$$

Normally there will be no ferromagnetic materials within the transmission line,[2] implying that the relative permeability is unity. But for example, the dielectric within the typical coaxial cable made with a foam polyethylene surrounding the center conductor has a typical relative permittivity of about 1.3 leading to

$$v = \frac{c}{\sqrt{(1.0)(1.3)}} = 0.88 \, c = v_f \, c$$

Thus for this particular cable the velocity factor v_f is 0.88. The velocity factor v_f is the fourth key parameter describing a transmission line.

By using {3-1} above

$$v = v_f \, c = \frac{\omega}{\beta} \quad or$$

$$\beta = \frac{\omega}{v_f \, c} = \frac{2\pi \, f}{v_f \, c} \qquad \{3-3\}$$

In addition, the source or signal frequency f, the wave's velocity v and wavelength λ of a wave in *free space* are also related by

$$\lambda = \frac{v}{f}$$

but within a transmission line, the wavelength is modified by the slower velocity of waves within the line and becomes λ_{TL}:

$$\lambda_{TL} = \frac{v_f \, c}{f}$$

Should we divide the numerator and the denominator of {3-3} by f, a new relationship becomes apparent:

---

[2]   This is not true when the transmission line is fabricated using iron, steel or nickel.

$$\beta = \frac{2\pi f}{v_f c} = \frac{2\pi}{\dfrac{v_f c}{f}}$$

$$\beta = \frac{2\pi}{\lambda_{TL}} \qquad\qquad \{3-4\}$$

Finally, as stated earlier R, L, C, G, β and α all have values per unit length. Obviously, then, these quantities must all be multiplied by a cable length d in meters for them to be useful in determining most circuit quantities involving transmission lines.

In summary, those parameters necessary to permit calculations in circuits involving transmission lines are listed in fig 3-1. When these items, the signal source frequency f, and the electrical properties of externally connected elements are known, almost any desired voltage, current, power, reflection coefficient, standing wave ratio (SWR), etc. can be computed. This includes voltage, current and impedance components at virtually any point along the transmission line itself. Given these parameters other quantities may also be calculated as needed.

Z_o Characteristic Impedance - ohms

α Attenuation - Nepers / meter

β Phase Shift - Radians / meter

v_f velocity factor (dimensionless)

d length - meters

Figure 3-1 Parameters Required to describe a Transmission Line

---

⌘ ⌘ ⌘

4 TRANSMISSION LINE SOLUTION & RELATIONSHIPS

The primary solution to the Transmission Line Equation is expressed in {4-1}:

$$Z_s = Z_o \frac{(Z_L + Z_o)\,e^{\gamma d} + (Z_L - Z_o)\,e^{-\gamma d}}{(Z_L + Z_o)\,e^{\gamma d} - (Z_L - Z_o)\,e^{-\gamma d}} \qquad \{4-1\}$$

(Johnson 1950: 96)

Where

Z_s = sending end impedance,
Z_L = receiving end or load impedance and
d = the line length measured from its receiving end.

This equation, used in several forms, has been programmed into the six Excel applications which are used to produce results for this book. By using {4-1} the following quantities can also be found:

$$k_s = \frac{(Z_L - Z_o)e^{-2\gamma d}}{Z_L + Z_o} \qquad \{4-2\}$$

$$k_r = \frac{Z_L - Z_o}{Z_L + Z_o} \qquad \{4-3\}$$

(Johnson 1950: 96)

Where:

k_s = voltage reflection coefficient at the cable's sending end, and
k_r = voltage reflection coefficient at the cable's receiving end

$$VSWR_r \quad = \quad \frac{1 + |k_r|}{1 - |k_r|} \qquad\qquad \{4-4\}$$

$$VSWR_s \quad = \quad \frac{1 + |k_s|}{1 - |k_s|} \qquad\qquad \{4-5\}$$

(Johnson 1950: 148)

where

$VSWR_r$ = voltage standing wave ratio at cable's receiving end, and
$VSWR_s$ = voltage standing wave ratio at the cable's sending end.

$$I_s \quad = \quad \frac{E_g}{Z_s + Z_g} \qquad\qquad \{4-6\}$$

Where

I_s = sending end current,
E_g = source voltage, and
Z_g = source impedance.

$$|E_s| \quad = \quad |I_s| \quad x \quad |Z_s| \qquad\qquad \{4-7\}$$

$$P_s \quad = \quad |I_s|^2 \quad x \quad Re[Z_s] \qquad\qquad \{4-8\}$$

Where

E_s = sending end voltage and
P_s = power supplied to sending end of the cable.

$$P_g \quad = \quad |I_s|^2 \quad x \quad Re\left[Z_g + Z_s\right] \qquad\qquad \{4-9\}$$

Where

P_g = total power produced by source

$$P_{SR} = |I_s|^2 \ x \ \text{Re}[Z_g]$$

$\{4-10\}$

Where

P_{SR} = power consumed by the source impedance.

$$I_r = \frac{2 I_s Z_o}{(Z_L + Z_o) e^{yd} - (Z_L - Z_o) e^{-yd}}$$

$\{4-11\}$

(Johnson 1950: 97)

Where

I_r = receiving end current.

$$P_r = |I_r|^2 \ x \ \text{Re}[Z_L]$$

$\{4-12\}$

Where

P_r = power consumed by the load impedance.

$$P_L = P_s - P_r$$

$\{4-13\}$

$$P_g = P_s + P_{SR}$$

$\{4-14\}$

Where

P_L = power consumed by the cable itself

$$E_x = \frac{I_s\left[(Z_s + Z_o)e^{-\gamma x} + (Z_s - Z_o)e^{\gamma x}\right]}{2} \qquad \{4-15\}$$

$$I_x = \frac{I_s\left[(Z_s + Z_o)e^{-\gamma x} - (Z_s - Z_o)e^{\gamma x}\right]}{2Z_o} \qquad \{4-16\}$$

(Johnson 1950: 95)

Where

x is the length to a point of the line measured from the cable's sending end,

E_x = the voltage at point x and
I_x = the current at point x.

The General application described in Chapter 6 computes all of the quantities listed in equations $\{4\text{-}1\}$ through $\{4\text{-}16\}$. Note that all quantities except d, x, $VSWR_s$, $VSWR_r$ and the power amounts are complex quantities.

⌘　⌘　⌘

5 USING THE TRANSMISSION
LINE APPLICATIONS

The two previous chapters have focused on the theory of transmission lines by describing mathematical relationships. With this chapter, the focus becomes less theoretical and more practical as actual problems and how to solve them are introduced. Clearly, the equations already listed can be used directly with function tables and any calculating device to obtain numerical results to proposed problems. However, specific programs have been developed and coded into Excel applications that perform the calculations accurately and with little effort, enabling the focus to be on the problems instead of the mechanics of their solution. (These applications are available on the CD.)

In the Introduction we indicated that this book would use Excel to perform the computer calculations. Obviously then, users who desire to use the applications described here must have access to a computer possessing Excel version 2003 or later.

The six specific applications are UnknownCable, Length, General, Reverse, ShuntReal, and StubMatch. Each one is designed to compute one or more specific unknown quantities in circuits that involve transmission lines; and each is described in upcoming chapters.

Each application has its own individual read-only application template file (a file containing input forms, programs and working sheets, etc. but lacking any problem-specific data). All these template files reside on the recommended CD and are named, respectively, UnknownCable.xls, Length.xls, General.xls, Reverse.xls, ShuntReal.xls and StubMatch.xls. It is recommended that each of these files be loaded onto the computer's main hard disk drive to use them. I also recommend that the files for each of the worked examples discussed in this book be loaded on to the main disc drive as well. This will make it easier to study the steps taken in their solution.

As these programs contain specific built-in macros, the computer's operating system may request that the user click an appropriate "activate" button for them to function. Windows® 7 users should create a new folder to receive the applications and example files from the CD. This new folder can be declared a "trusted location" from which all the applications can be opened and run without having to "enable content" each time a program is used. From that time forward, no questionable data or programs from other sources or applications should ever be stored in that location.

The data handling program statements, formulas, and other code for each application exist on unseen computation sheets within each application file.

To use an application, the user opens the desired template file into an open Excel environment in the normal way.

Each application file should open showing the InOut1 sheet as indicated by the tab at the bottom left of the screen. If InOut1 is not the sheet displayed, merely select the InOut1 tab.

The StubMatch Application

Data for:	StubMatch.xls
Input Data	
Z_o -- T. Line Char. Imped. Ohms	0
Z_L -- Rec. end load Imped. Ohms	0
β -- T. Line ph. shift radians/m	0
Output Results	
d1 distance to junctn from load	0.000000
d1 in wavelengths	0.000000
d2 length of stub	0.000000
d2 in wavelengths	0.000000

Figure 5-1 A Typical Application InOut1 Sheet

The key data sheet or input form for each application is called InOut1. When open, the application's name appears at the top of its InOut1 sheet and the second line lists the sheet's file name where the program and sheet are stored. A sample copy of an empty InOut1 template sheet for the StubMatch application is shown in fig. 5-1.

When an application has been previously executed, its InOut1 sheet will contain input data in its top section and calculated results in its lower portion when it opens.

The General application also has a second data sheet labeled Out2, which may be displayed by clicking the Out2 tab at the lower left of the screen when that application is open. When the General application is executed and the x function and its limits, Startx and Stopx are used, Out2 will contain tables of calculated data and graphs plotted from this data. See chapter 6 for a discussion of the x function, Startx and Stopx.

In those instances where the x function is not used, Out2 can be ignored, but when the x function is used a table of x, E_x, I_x and the impedance components of the E_x and I_x ratio are produced by the program. In addition, graphs of these four quantities versus x appear on the Out2 sheet. The Template file Out2 sheet for the General application plots contains default titles and scales. The user may add or change the labels and the graph nomenclature as desired. Consult an Excel Manual for instructions.

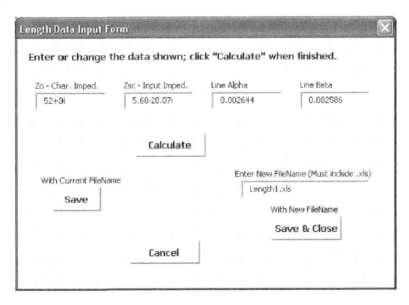

Figure 5-2   A Typical Application Data Input Form

When any application file is opened, a button labeled **Enter/Change Data** appears on the InOut1 sheet that is displayed. Click this button and a data input form will appear. Fig. 5-2 shows an example of such a data input form. If the user has opened an application file and clicked the button of a previously created file, the data numbers previously entered into that file will appear within the data entry form boxes and on the InOut1 sheet when it opens. For a new file (i.e., when a template file is opened for the first time), the boxes of the data input form will all contain zero while the spaces on the InOut1 sheet will contain zeros, blanks, or #NUM!, all indicating that the template file has not been exercised. In any case, click the desired box on the input form to enter or change the data in that box.

The programs and formulas employed in these applications are based on data with specific units. The computed results will be in error unless each data value is entered using the correct units. The units required by these applications are as follows:

- All lengths are in meters (for conversions see Appendix A).
- All voltages are in volts.
- All impedances are in ohms.
- All attenuation factors α are in nepers per meter (see manufacturer's data in chapter 6 or Appendix A).
- All phase shift factors β are in radians per meter.

Important! The data cell boxes expect complex quantities for all voltages and impedances!

When writing complex electrical expressions, i and or j are often used to indicate the imaginary terms; and such terms may be written with the i or j at the beginning of the term or at its end. Excel requires that the imaginary operator be i and that it be entered last. That is, in Excel, complex quantities must have the following form: a + bi or a − bi. The i must be entered.

When finished entering data, click the **Calculate** button on the input form. When this is done, all of the data in the form are transferred to the top of the application's InOut1 sheet and to the application's calculation engine.[3] The **Calculate** button will remain depressed while the

---

[3]   The word 'engine' refers to that part of the application which performs the calculations.

calculations continue but will return to its normal up position when they finish. The application's calculated results are then written to the lower part of the InOut1 sheet. Assuming that the user's computer speed is typical, these actions are immediate except for the StubMatch Application, which may require three or four seconds.

After clicking **Calculate** on the input form, the user may change one or more data entries and click **Calculate** again, or may want to print the results. A printed copy can be obtained via the Excel menu in the usual way, or the user may want to save the results for future reference.

To save new results, enter a name different from the template file name (this name must contain the .xls suffix) into the box above the **Save & Close** button, and then click **Save & Close**. These actions save the file. During this process, Excel will prompt, "Do you wish to save 'the old file'?" (The prompt will list the actual name of the old file.) Because the application template files are all of the read-only type, a "yes" answer will save the template file under a copy name. A "no" answer closes the old file, unchanged.

There may be times when the user desires to make modifications to an old or previously saved non template file and then resave the modified file under the old name. To do this, open the old file, click **Enter/ Change Data** on the InOut1 sheet, make the changes on the form, click **Calculate** and click **Save**.

Important Notes

The data input forms for the six applications are designed to remain on the screen when **Calculate** is clicked. In this way, the user is able to quickly modify one or more data entries and click **Calculate** after each data entry, repeatedly, without having to recall the form before entering the changed data, and clicking **Calculate**.

Sometimes users may want to copy results from the application's InOut1 or Out2 sheets and paste them to other spreadsheets or other program sheets. This may be done by opening the receiving sheet. This means that more than one sheet is in computer memory simultaneously. Thus, when an application's sheet is open, results may be copied from it. Then the second sheet may be selected and the copied results pasted to it.

This means that any sheet, not just the one linked to the input form, can become active and the data input form, which remains linked to the open application, will remain on the screen. Should the form obstruct areas of interest on the screen, it can be removed by clicking **Cancel** on the form; or the form may be moved by clicking on the form's blue band while holding the mouse button and moving the form out of the way.

This form display flexibility, however, can cause problems. For example, should any button on a displayed form, other than **Cancel**, be clicked while the active sheet on the screen under the form is *not* the application that is linked to that form will cause an error! Recovery from this error will cause the loss of data that may have been entered into the form after a previous **Save** or **Save & Close** operation. This happens because getting back to the point where the error was produced requires clearing the error notice window and restarting the application being used. Restarting that application will restore the data last saved in the file to the form and to the InOut1 and Out2 sheets of the application.

Difficulties caused by this error can be minimized by remembering to save results frequently and before adding additional program sheets to memory or before using the copy/paste routine.

In Case of Difficulty When Using the Programs

The programs have been tested using many different data values and under various conditions in a concerted effort to remove odd behaviors and program bugs. The user can be confident of program operation and results, but, as with all computer programs, unforeseen events are possible.

If after opening an application and clicking the **Enter/Change Data** button and nothing appears to happen, look for the security note about center screen, and click "Enable the content" or put the application files into a trusted location as described earlier, and open them from that location.

During the many tests of these applications, the most frequent cause for errors was data entry mistakes. This should be the first area checked when unexpected results occur. Check that an "i" has been appropriately entered where required, and make certain that commas have not been entered in place of decimal points. Verify that minus signs are "–" and not "_" instead. It is easy to enter two decimal points into a real or imaginary quantity, only one per real or imaginary quantity is allowed. All of these errors will prevent proper results.

Finally, if after trying a number of entries or re-entries without success, close the current file without saving it, go to the trusted site, select the proper file again, reload it, and make a fresh start.

If you suspect that the application file on your hard disc drive is corrupted, simply reload it from the CD and begin again.

Final Notes on Application Program Use

The forms in the six applications operate in the same way.

In the chapters that follow, specific characteristics of each of the applications are discussed, and the applications will be used to solve problems. These steps demonstrate how to use the programmed applications and illustrate the behavior of transmission lines by solving meaningful problems.

⌘   ⌘   ⌘

6 THE GENERAL APPLICATION

As the name implies, the General application calculates numerous quantities associated with transmission lines in use. Specifically, this application computes all of the quantities described in equations {4-1} through {4-16}. To use this application, open the General.xls file by following the instructions given in chapter 5.

Some functions within the General application deserve additional explanation. As mentioned in Chapter 4, equations {4-15} and {4-16} provide, a so called x function for the application. This function uses the transmission line situation already characterized by Z_o, Z_L, α, β, and d and provides the capability to compute point values of voltage, current, resistance and reactance for any value of x along the line's length. In all cases, x is the distance in meters measured from the cable's sending end.

When the General application employs the x function, it produces a table of fifty values for current, voltage, resistance and reactance as a function of x, equally spaced between the distances Startx and Stopx, which were entered into the General Application Data-Input Form. The data in this table are used to plot two graphs. These calculated results in table and graphical form appear on the Out2 sheet.

Startx and Stopx can be set to produce tables and plots of the four quantities over any range of x values located anywhere along the cable's length or to include its entire length. Startx should always be less than or equal to Stopx and both Startx and Stopx must be less than or equal to d. Neither Startx nor Stopx can be less than zero. Data for the entire cable length are plotted when Startx is set equal to zero and Stopx is set equal to d.

Example 6-1[4].

Given a telephone transmission line 100 miles long, with Z_o = 685-j92 ohms, α = 0.00497 nepers/mile, and β = 0.0352 radians/mile at 1000 Hz. The line is terminated with Z_L = 2000+j0 ohms. The generator has an emf of 10 volts rms and an internal impedance of 700 +j0 ohms. Find the sending-end impedance, the sending-end current, voltage and power and the receiving-end voltage, current and power. (Johnson 1950: 97)

Solution: Because the line length d, α and β all use the same length dimensions of miles, no length conversions are necessary and the quantities stated in miles may be entered directly into the data input form along with all of the other data listed in the previous paragraph. To see how the data have been entered, open Example 6-1.xls file on the recommended CD and click the button.

The output results are shown in Example 6-1.xls. As can be seen, all of the quantities required by the problem and more have been calculated and are shown on the sheet. The results that Johnson obtained have also been included for comparison. Because Johnson's work was published in 1950, his results were almost certainly not obtained by using a computer; and as the differences between the two sets of results range from -0.58% and 0.12%, his method of calculation probably employed extrapolated function tables and a mechanical calculator.

Example 6-1 Comments
The reflection coefficients and voltage standing wave ratios (VSWRs) are useful in ascertaining the power transfer effectiveness of the transmission line and its associated components. In an ideal transmission line setup there will be no reflections at either line end, but the results for this problem show that the reflection coefficient at the load is 49% and is 18% at the sending end. (See Example 6-1.xls.) With reflection values like these there will always be higher VSWRs than desired. Note that the calculated VSWRs are 1.44 at the sending end and 2.92 at the load end.

---

[4]   All examples are included on the recommended CD. Each is stored in its solved form.

The General Application			
Data for:	**Example 6-1.xls**		
Input Data			
Z_0 -- T. Line Char. Imped. Ohms	685-92i		
Z_L -- Rec. end load Imped.Ohms	2000+0i		
d -- T. Line length meters	100		
α -- T. Line atten. nepers/m	0.00497		
β -- T. Line ph. shift radians/m	0.0352		
V_g -- Send.end source voltage	10+0i		
Z_g -- Send.end source imped.	700+0i		
Startx	0		
Stopx	100		
		Results	
		Produced by	Dif.
Output Results		Johnson	%
K_r -- Refl. Coef. at rec. end	0.488010913188418+5.0985848794538E-002i		
abs| K_r |	0.490667105		
K_s -- Refl. Coed. at send end	0.14426399534468-0.11028954810988i		
abs| K_s |	0.181592634		
$VSWR_s$ -- VSWR at send. end	1.443770772		
$VSWR_r$ -- VSWR at rec. end	2.926704952		
Z_s -- send. end imped.	862.542895450651-322.471057300864i	861-325i	0.06
Y_s -- send . end Admittance	1.01718823009328E-003+3.80286900236768E-004i		
I_s -- send. end current	6.13838452108495E-003+1.2668140838864E-003i		
abs| I_s |	0.006267741	0.00626	0.12
I_r -- rec. end current	-2.09225126640775E-003+8.90320662074085E-004i		
abs| I_r |	0.002273804	0.00228	-0.27
E_s -- send. end voltage	5.70313083524054-0.886769858720482i		
abs| E_s |	5.771660256	5.75	0.38
E_r -- rec. end voltage	-4.1845025328155+1.78064132414817i		
abs| E_r |	4.547608709	4.56	-0.27
P_g -- Total gen/source Pwr.	0.063998243		
P_{sr} -- Pwr used by source	0.027499208		
P_s -- Pwr supplied to T. Line	0.033884637	0.0338	0.25
P_r -- Pwr consumed by load	0.010340372	0.104	-0.58
P_L -- Pwr consumed by T. Line	0.023544265		

As most modern transmitters begin limiting their output power when they work into a load with a VSWR of 2.0 and higher, one must conclude that the situation described by this problem is marginal at best.

The standing waves along the transmission line can be viewed by studying the plots that come from the results of Example 6-1.xls. Fig. 6-1 shows the standing waves in voltage and in current while fig. 6-2 shows how the impedance components vary along the length of the transmission line.

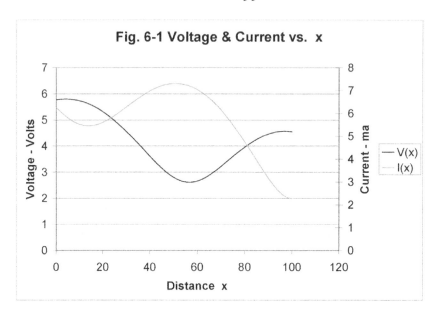

Fig. 6-1 Voltage & Current vs. x

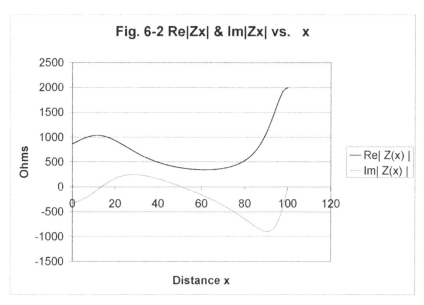

Fig. 6-2 Re|Zx| & Im|Zx| vs. x

By inspection it is clear that standing waves exist along the transmission line because the impedances at neither end are matched to the line. The calculated results confirm that reflection coefficients and VSWRs are not ideal, but fig. 6-1 shows us something more subtle.

As the line has a finite attenuation, the VSWR at the load end is greater than it is at the source end. This is apparent in fig. 6-1 because the peak to peak variations of voltage and current near the load end of the line are greater than they are near the source end. This is confirmed by the calculated results for source VSWR and load VSWR as described above. Because there is non-negligible attenuation, the absolute value of voltage and current declines as the distance from the generator increases.

These phenomena are even more apparent should the length in this example be longer. See example 6-2.

The first thing noted from fig. 6-2 is that as x reaches the load end, at 100 miles, the point impedance values become equal to the components of the load impedance as should be expected. Note that the changing values for Z_x as x varies also suggests the existence of standing waves, a fact already made clear by fig. 6-1.

Example 6-2.

Repeat example 6-1 but change only d to 300 miles instead of 100 miles.

Solution: Example 6-2.xls shows the Input and calculated Output results; fig. 6-3 and fig. 6-4 show the plots for example 6-2.

Example 6-2 Comments

Of course, the act of lengthening the transmission line creates a problem entirely different from example 6-1, resulting, in general, in different calculated values. For example the Example 6-2 source end VSWR is only 1.05 compared to 1.44 for Example 6-1.xls; however, the load end VSWR is the same as it was for Example 6-1.xls, i.e. $VSWR_r = 2.92$.

Again, note that the peak to peak swings in voltage and current of fig. 6-3 are greater near the load, indicating the larger VSWR at the load; at the same time the magnitudes of current and voltage decline as x increases from the source. This agrees with our intuition, i.e., that signal strengths will decline as the distance from the source increases.

The General Application

Data for:	Example 6-2.xls
Input Data	
Z_o -- T. Line Char. Imped. Ohms	685-92i
Z_L -- Rec. end load Imped.Ohms	2000+0i
d -- T. Line length meters	300
α -- T. Line atten. nepers/m	0.00497
β -- T. Line ph. shift radians/m	0.0352
V_g -- Send.end source voltage	10+0i
Z_g -- Send.end source imped.	700+0i
Startx	0
Stopx	300
Output Results	
K_r -- Refl. Coef. at rec. end	0.488010913188418+5.0985848794538E-002i
abs| K_r |	0.490667105
K_s -- Refl. Coed. at send end	-1.39525947688819E-002-2.05905373146841E-002i
abs| K_s |	0.024872578
$VSWR_s$ -- VSWR at send. end	1.051014005
$VSWR_r$ -- VSWR at rec. end	2.926704952
Z_s -- send. end imped.	661.907431093292-116.8199673335468i
Y_s -- send . end Admittance	1.46514785289735E-003+2.58583778149146E-004i
I_s -- send. end current	7.28901296328633E-003+6.25227704055743E-004i
abs| I_s |	0.007315779
I_r -- rec. end current	-3.26633125438696E-004+7.69168310747971E-004i
abs| I_r |	0.000835649
E_s -- send. end voltage	4.89769092569958-0.437659392839018i
abs| E_s |	4.917206742
E_r -- rec. end voltage	-0.653266250877392+1.53833662149594i
abs| E_r |	1.671297806
P_g -- Total gen/source Pwr.	0.07342643
P_{sr} -- Pwr used by source	0.037464434
P_s -- Pwr supplied to T. Line	0.035425696
P_r -- Pwr consumed by load	0.001396618
P_L -- Pwr consumed by T. Line	0.034029078

Note also that when $x = 300$ miles, the impedance components in fig. 6-4 again equal the real and imaginary load impedance components connected to the transmission line.

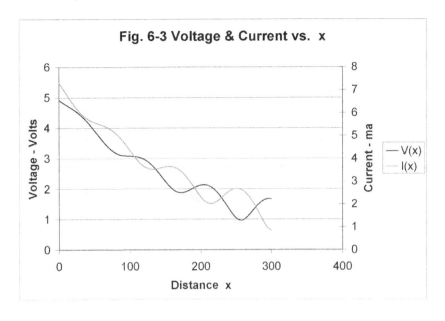

Fig. 6-3 Voltage & Current vs. x

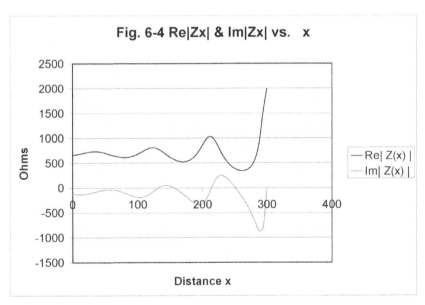

Fig. 6-4 Re|Zx| & Im|Zx| vs. x

Example 6-3.

Keep all parameters as they were in example 6-2 except that the transmission line is assumed to be lossless, i.e., the attenuation factor α is set equal to zero.

Solution: The results for this example are shown in Example 6-3.xls, fig. 6-5 and fig. 6-6.

The General Application

Data for:	Example 6-3.xls
Input Data	
Z_o -- T. Line Char. Imped. Ohms	685-92i
Z_L -- Rec. end load Imped.Ohms	2000+0i
d -- T. Line length meters	300
α -- T. Line atten. nepers/m	0
β -- T. Line ph. shift radians/m	0.0352
V_g -- Send.end source voltage	10+0i
Z_g -- Send.end source imped.	700+0i
Startx	0
Stopx	300
Output Results	
K_r -- Refl. Coef. at rec. end	0.488010913188418+5.0985848794538E-002i
abs| K_r |	0.490667105
K_s -- Refl. Coed. at send end	-0.275246069542175-0.4061943000021403i
abs| K_s |	0.490667105
$VSWR_s$ -- VSWR at send. end	2.926704952
$VSWR_r$ -- VSWR at rec. end	2.926704952
Z_s -- send. end imped.	248.622204803843-349.665362800874i
Y_s -- send . end Admittance	1.35062872403162E-003+1.89954104529966E-003i
I_s -- send. end current	9.28065595129265E-003+3.42088126738491E-003i
abs| I_s |	0.009891057
I_r -- rec. end current	-1.41433003627876E-003+3.5296789095571E-003i
abs| I_r |	0.003802494
E_s -- send. end voltage	3.50354083409514-2.39461688716943i
abs| E_s |	4.243699849
E_r -- rec. end voltage	-2.82866007255752+7.0593578191142i
abs| E_r |	7.604988535
P_g -- Total gen/source Pwr.	0.105416044
P_{sr} -- Pwr used by source	0.068483102
P_s -- Pwr supplied to T. Line	0.024323457
P_r -- Pwr consumed by load	0.028917925
P_L -- Pwr consumed by T. Line	0.004594468

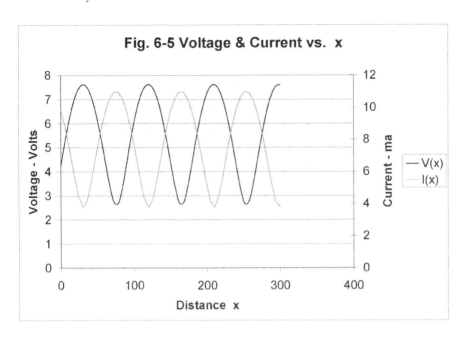

Fig. 6-5 Voltage & Current vs. x

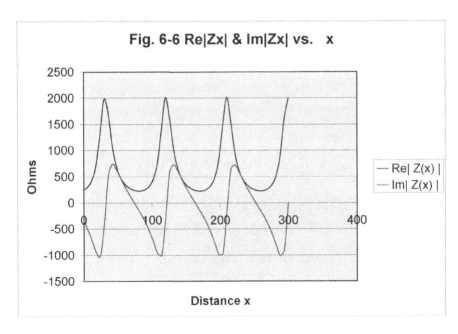

Fig. 6-6 Re|Zx| & Im|Zx| vs. x

Example 6-3 Comments

Note first how different the curves are when the line is lossless as compared with the case where the line has losses. Not only do the magnitudes vary more widely, but the periodicity in the lossless case is clear. As expected, since the attenuation is zero, the standing waves do not die out along the line and the peak to peak swing of the standing waves are the same along the line's entire length, i.e. the VSWR at the load end and VSWR at the source end both = 2.92.

One can now clearly see the standing waves. Because standing waves are the vector sum of waves travelling away from the source combining with the same frequency waves reflected at the line ends, a relationship between the standing wave and the signal wave on the line can now be observed. To illustrate this point, let us compute the wavelength of the signal carried by the transmission line for this example.

Because the length was specified in miles, the free space wavelength λ will be expressed in miles as well (If this equation confuses you, see Appendix A.)

$$\lambda = \frac{c}{f}$$

$$= \frac{\left(2.9979 \times 10^8 \, m/\sec\right)}{1000 \, hz} \, x \, \frac{1 \, in}{2.54 \, cm} \, x \frac{100 \, cm}{m} \, x \, \frac{1 \, ft}{12 \, in} \, x \frac{1 \, mile}{5280 \, ft}$$

$$\lambda = 186.280 \, miles$$

and from {3-4}:

$$\beta = \frac{2\pi f}{v_f c} = \frac{2\pi}{\lambda_{TL}}$$

Because β was specified in the problem as 0.0352 radians/mile:

$$\lambda_{TL} = \frac{2\pi}{0.0352} = 178.5 \, miles$$

With the wavelength in free space and the wavelength within the cable both known, the v_f for the transmission line can also be computed:

$$v_f = \frac{\lambda_{TL}}{\lambda} = \frac{178.5}{186.28} = 0.9582$$

As wave phenomena within a transmission line with standing waves repeat every half wavelength, similar magnitude behaviors of interest should repeatedly occur along the line; in this case, each occurrence should be separated by an $x = 178.5$ miles$/2 \sim= 90$ miles. Thus approximately 90 miles should separate similar peaks and positive going zero crossings, etc., on the curves shown in fig. 6-5 and fig. 6-6. Study of these figures confirms this fact.

Cheng (1989) also describes an interesting relation for *lossless* transmission lines. For the case where the line has low losses, the effect is similar. To use his analysis, start at the load end of the transmission line; at that point the impedance curves will indicate that the point impedance components are equal to the actual load impedance values; next, move toward the source by a half wavelength, 90 miles in this case; the impedance components at this new point will again equal the actual load impedance components. This result will repeat for each half wave-length as one moves toward the source. Check this in fig. 6-6.

Example 6-4.

A 105 foot length of Belden 7810A coaxial cable with $Z_o = 50 + 0i$ ohms is used to connect a transmitter to an antenna. The transmitter is operated at 14 Megahertz (MHz) and has a nominal output impedance of $50 + j0$ ohms. At 14 MHz the antenna has an input impedance of $100-j200$ ohms and is thus not matched to the line. From the manufacturer's tables v_f for 7810A was found to be 0.86 and its attenuation is 0.48 decibels (db)/100 ft at 14 MHz. Assume that the transmitter will supply its maximum output of 100 watts to the line. Find the actual power supplied to the antenna.

Solution: First, take care of the units conversions: See Appendix A:

$$d = 105 \; feet \; \; x \; \frac{12 \, in}{ft} \; x \; \frac{2.54 \, cm}{in} \; x \; \frac{1 \, meter}{100 \, cm}$$

$$= 32.00 \; meters$$

$$\alpha \; = \; 0.48 \,_{db/100\,ft} \; \; x \; \frac{0.0037772065 \; 2 \, neper \; \; ft}{db \; meter}$$

$$= \; 0.001813 \; \frac{nepers}{meter}$$

Then from {3-3}:

$$\beta = \frac{2\pi f}{v_f c} = \frac{(2\pi)(14 \times 10^6)}{(0.86)(2.99792 \times 10^8)}$$

$$= 0.34118 \frac{radians}{meter}$$

Despite the fact that the transmitter's output impedance nominally equals the line Z_o, it is unlikely that the input impedance of the line, connected with the antenna load as specified, will match the transmitter output impedance because of the mismatch of the line at the antenna; but we desire to eventually match these impedances. Consequently assume there is such a match for the purposes of finding the source current and source voltage necessary to provide the required output power. See fig. 6-7.

The power into the line being 100 watts, then the transmission line input voltage is calculated as follows:

$$P_o = \frac{V_o^2}{Re|Z_o|}$$

Where V_o is the transmission line input voltage; solving for V_o we get:

$$V_o = \sqrt{(100 \; watts)(50 \; ohms)} = 70.7 \; volts \; rms$$

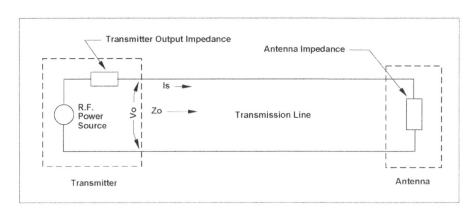

Figure 6-7 Model of Transmitter, Line and Load

Because the transmission line, under our assumption, is matched to the transmitter, the voltage drop across the transmitter output impedance will also equal V_o (Remember the same current flows in the line and in the transmitter output impedance); thus the generator (transmitter) voltage must equal $2V_o$, i.e. $V_{gen} = 141.4$ volts rms. Again, see fig. 6-7.

Having obtained the necessary input parameters, open General. xls, click the button, enter the data, click **Calculate,** and save the file as Example 6-4.xls. Voltage and current plots from Example 6-4.xls are shown in fig. 6-8 and the impedance components are shown in fig. 6-9.

Example 6-4 Comments

As can be seen from Example 6-4.xls only 28 of the intended 100 watts are actually absorbed by the antenna, but even this low number can only be produced when the transmitter is severely overloaded! See P_g and P_{sr} in Example 6-4.xls.

An inspection of the problem specifications immediately tells us that there would be standing waves on the transmission line, but the Example 6-4.xls results demonstrate that the mismatches are serious.

The General Application	
Data for:	**Example 6-4.xls**
Input Data	
Z_o -- T. Line Char. Imped. Ohms	50+0i
Z_L -- Rec. end load Imped. Ohms	100-200i
d -- T. Line length meters	32
α -- T. Line atten. nepers/m	0.001813
β -- T. Line ph. shift radians/m	0.34118
V_g -- Send. end source voltage	141.4
Z_g -- Send. end source imped.	50+0i
Startx	0
Stopx	32
Output Results	
K_r -- Refl. Coef. at rec. end	0.76-0.32i
abs| K_r |	0.824621125
K_s -- Refl. Coed. at send end	-0.712727113835852+0.176603842372525i
abs| K_s |	0.734281183
$VSWR_s$ -- VSWR at send. end	6.526753369
$VSWR_r$ -- VSWR at rec. end	10.40388203
Z_s -- send. end imped.	7.77217088085025+5.95704200047301i
Y_s -- send . end Admittance	8.10504595741604E-002-6.2121767468542E-002i
I_s -- send. end current	2.42179613896389-0.2497178331147 52i
abs| I_s |	2.434636633
I_r -- rec. end current	-0.450577344845004+0.286071593618408i
abs| I_r |	0.533719871
E_s -- send. end voltage	20.3101930518055+12.4858916557376i
abs| E_s |	23.84117095
E_r -- rec. end voltage	12.1565842391812+118.722628330842i
abs| E_r |	119.3433912
P_g -- Total gen/source Pwr.	346.0828924
P_{st} -- Pwr used by source	296.3727767
P_s -- Pwr supplied to T. Line	46.06919731
P_r -- Pwr consumed by load	28.48569004
P_L -- Pwr consumed by T. Line	17.58350727

The 346 watt total transmitter power number clearly indicates that the transmitter would be in power limit mode. Should the transmitter not be equipped with this feature, the transmitter output voltage must be manually reduced to prevent damage to the transmitter output stage. This will be discussed later.

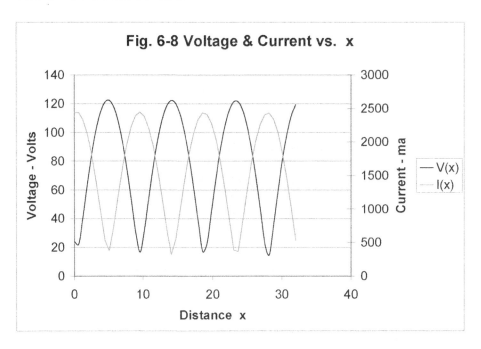

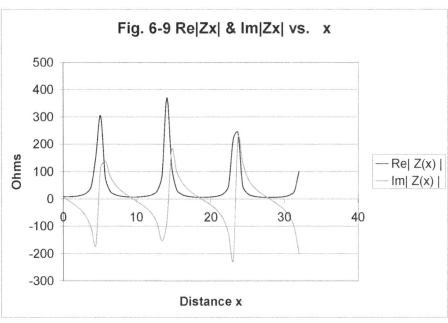

The problem specifications indicate that the transmitter is expected to supply 100 watts to the transmission line input. This implies that the transmitter should be capable of producing 200 watts: 100 watts to the line and an additional 100 watts of loss within the transmitter output stage under optimum conditions.

Transmitters equipped with the power limiting feature contain circuitry which senses excessive output stage power dissipation and automatically begin to reduce power should the overload persist. This implies that the sensors are measuring what is labeled in our example as P_{sr}, power used by the Source, i.e. the 296.4 watts shown in Example 6-4. xls. Thus it is clear that V_g must be reduced such that P_{sr} is less than or equal to the transmitter output rating of 100 watts.

By repeatedly adjusting V_g and clicking **Calculate** on the input form of Example 6-4.xls, we arrive at $P_{sr} = 100$ watts when $V_g = 82.135 + 0i$ volts. Having reached this value, save the modified Example 6-4.xls file under the new name of Example 6-4a.xls. The new graphical results are shown in fig. 6-10 and fig. 6-11.

Example 6-4a Comments

Reducing the transmitter output voltage does reduce the power levels, currents and voltages, but as expected, the transmission line/antenna impedance situation has not changed from Example 6-4.xls. Note that the VSWR calculations are the same in Example 6-4.xls and Example 6-4a.xls. Note also that when the output stage is at its rated level, only 9.6 watts are delivered to the antenna!

As in Example 6-1.xls, note that there are slight declines in point currents and voltages as well as increases in peak to peak voltages, currents, and impedance components as x increases. See fig. 6-10 and fig. 6-11. This should be expected because, while the cable losses are small, they are finite; and the computed VSWRs at the two cable ends are not equal.

Again, the impedance components equal the load impedance values when x is equal to the cable length. See fig. 6-11. Note also the strong periodicity in fig. 6-10 and fig. 6-11.

The wavelength of the signal carried by the cable from {3-4} is:

$$\lambda_{TL} = \frac{2\pi}{\beta} = \frac{2\pi}{0.34118} = 18.42 \; meters$$

Therefore peaks, positive-going zeros etc. should repeat about every 18.42/2 = 9.2 meters; and by inspection of fig. 6-10 and fig. 6-11, this is found to be the case.

The General Application

Data for:	Example 6-4a.xls
Input Data	
Z_o -- T. Line Char. Imped. Ohms	50+0i
Z_L -- Rec. end load Imped.Ohms	100-200i
d -- T. Line length meters	32
α -- T. Line atten. nepers/m	0.001813
β -- T. Line ph. shift radians/m	0.34118
V_g -- Send.end source voltage	82.135+0i
Z_g -- Send.end source imped.	50+0i
Startx	0
Stopx	32
Output Results	
K_r -- Refl. Coef. at rec. end	0.76-0.32i
abs| K_r |	0.824621125
K_s -- Refl. Coed. at send end	-0.712727113835852+0.176603842372525i
abs| K_s |	0.734281183
$VSWR_s$ -- VSWR at send. end	6.526753369
$VSWR_r$ -- VSWR at rec. end	10.40388203
Z_s -- send. end imped.	7.77217088085025+5.95704200473301i
Y_s -- send . end Admittance	8.10504595741604E-002-6.2121767468542E-002i
I_s -- send. end current	1.40674841494908-0.145053565932674i
abs| I_s |	1.414207071
I_r -- rec. end current	-0.261726804942324+0.166170370168657i
abs| I_r |	0.310021793
E_s -- send. end voltage	11.7975792525463+7.252678296663374i
abs| E_s |	13.84861793
E_r -- rec. end voltage	7.061393539499+68.9623980053305i
abs| E_r |	69.32298044
P_g -- Total gen/source Pwr.	116.7717626
P_{sr} -- Pwr used by source	99.999082
P_s -- Pwr supplied to T. Line	15.54419906
P_r -- Pwr consumed by load	9.611351235
P_L -- Pwr consumed by T. Line	5.93284783

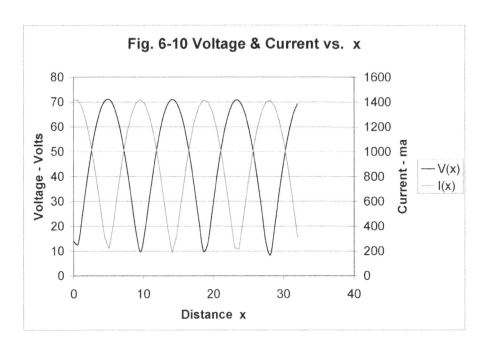

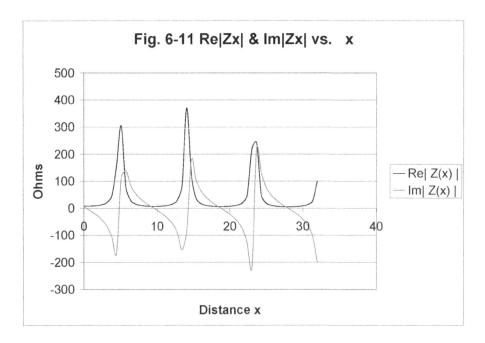

Transmitter Tuners

Faced with this situation, many amateur radio operators will most likely get out the equipment catalogs and look for some magical piece of hardware which will reduce the feed line SWR. Several companies offer antenna tuners, some with their own built-in SWR meter, and a number of higher-priced transceivers include antenna tuners and SWR meters, implying to buyers that they fix such problems. But what can a tuner really do for us in this case?

A tuner's purpose is to maximize power transfer to its load. It does this by matching its input impedance to the output impedance of its upstream energy supplier, the transmitter, and adjusting its output impedance to enable maximum power transfer to whatever load is connected to its output terminal. See fig. 6-12.

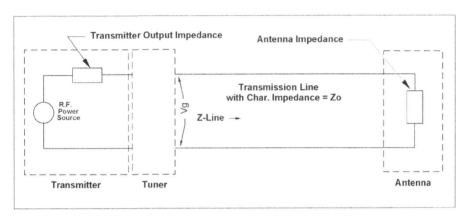

Figure 6-12 Model of System Using a Tuner

To calculate the effect of using a tuner at the transmitter's output, let us simulate a transmitter/tuner combination connected to our transmission line/antenna by using the General application. This requires us to recognize that the match that binds the tuner input to the transmitter output has the effect of creating the equivalent of a "new piece of hardware" that supplies power, as the transmitter alone did, but this "new piece of hardware" also adds the ability to adjust its output impedances over a significantly wider range than the transmitter alone could. Of course, the new equivalent hardware combination retains the ability to supply a variable output voltage. The task then is to determine the values of the tuner output impedance and output voltage that should be used on the General application input form in order to carry out the simulation.

It is known from circuit theory that maximum power transfer occurs when the load impedance is the complex conjugate of the impedance seen when looking back toward the generator.

We need to be clear, however: "…a *line* is *matched* when the load impedance is *equal* to the characteristic impedance (not the complex conjugate of the characteristic impedance) of the line." (Cheng 1989: 449) However, ours is not a case of line impedance matching; it is instead a case of maximum power transfer. (Remember that the line is not matched to its load impedance.) Thus the tuner output impedance should be the conjugate of the impedance seen when looking into the transmission line leading to the antenna, i.e. the conjugate of the impedance labeled as Z-Line in fig. 6-12.

The value of Z-Line has been computed in all versions of example 6-4 up to this point and is 7.7721708 +5.957042i ohms, making its complex conjugate = 7.7721708 -5.957042i ohms. Next, modify Example 6-4a.xls by entering this conjugate number into Z_g on its input data form; and then enter V_g = 55.756+0i volts into the same input form. (This voltage was determined experimentally as was done previously in order to limit the output power dissipation P_{sr} to 100 watts, this time for this "new piece of hardware" using this different output impedance.) This new file is saved as Example 6-4b.xls; its file produces the graphs shown in fig. 6-13 and fig. 6-14.

Example 6-4b Comments

As the Example 6-4b.xls results indicate[5], the use of an antenna tuner increases the percentage of power absorbed by the antenna from 9.6% to nearly 62%, a clear improvement. Note however, the VSWRs calculated for this case are the same as they were for Example 6-4.xls and Example 6-4a.xls.

These calculations have just shown that even by providing a perfect match at the transmitter/tuner interface and a conjugate match at the tuner/cable interface, nothing related to the impedance of the transmission line has changed. This is what we should expect; after all, a tuner cannot alter any impedance beyond its output connector. The transmission

---

[5] Note that P_s and P_{sr} are almost identical; this occurs because the application of the complex conjugate rule makes the real part of the source impedance equal the real part of the load impedance.

line/antenna impedance relationship remains exactly as it was before: badly mismatched with large standing waves and large reflection coefficients. See fig. 6-13 and fig. 6-14. All that the tuner accomplished was to improve the match at the transmitter end of the transmission line only; the antenna end of the line remains as it was, unmatched.

The General Application

Data for:	Example 6-4b.xls
Input Data	
Z_o -- T. Line Char. Imped. Ohms	50+0i
Z_L -- Rec. end load Imped.Ohms	100-200i
d -- T. Line length meters	32
α -- T. Line atten. nepers/m	0.001813
β -- T. Line ph. shift radians/m	0.34118
V_g -- Send.end source voltage	55.756+0i
Z_g -- Send.end source imped.	7.7721708-5.957042i
Startx	0
Stopx	32
Output Results	
K_r -- Refl. Coef. at rec. end	0.76-0.32i
abs$\mid K_r \mid$	0.824621125
K_s -- Refl. Coed. at send end	-0.712727113835852+0.176603842372525i
abs$\mid K_s \mid$	0.734281183
$VSWR_s$ -- VSWR at send. end	6.526753369
$VSWR_r$ -- VSWR at rec. end	10.40388203
Z_s -- send. end imped.	7.77217088085025+5.95704200473011i
Y_s -- send . end Admittance	8.10504595741604E-002-6.2121767468542E-002i
I_s -- send. end current	3.5869000530713-1.09148369313793E-009i
abs$\mid I_s \mid$	3.586900053
I_r -- rec. end current	-0.703554257602094+0.351152665765103i
abs$\mid I_r \mid$	0.786318503
E_s -- send. end voltage	27.878000151503+21.3673142744312i
abs$\mid E_s \mid$	35.12470657
E_r -- rec. end voltage	-0.124892607188798+175.826118096929i
abs$\mid E_r \mid$	175.8261625
P_g -- Total gen/source Pwr.	199.9911994
P_{sr} -- Pwr used by source	99.99559916
P_s -- Pwr supplied to T. Line	99.9956002
P_r -- Pwr consumed by load	61.82967881
P_L -- Pwr consumed by T. Line	38.16592139

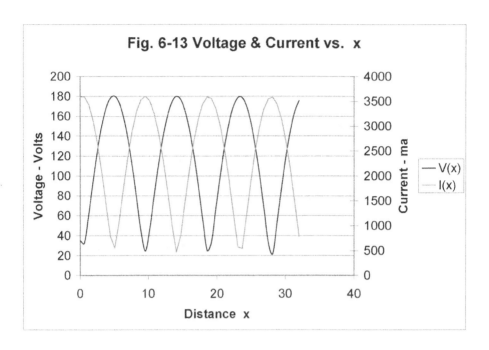

Fig. 6-13 Voltage & Current vs. x

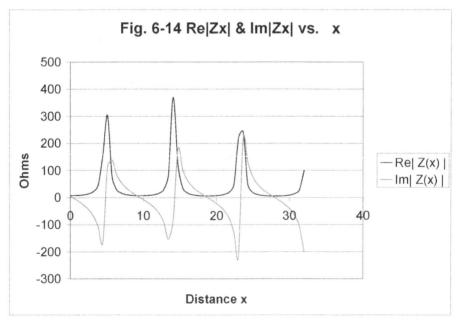

Fig. 6-14 Re|Zx| & Im|Zx| vs. x

Another point can also be made here. There are some radio operators who believe that transmission lines must have VSWR values near 1.0 for even minimal power to be available at the antenna. While it is true that a system possessing a high VSWR value is a sign of a system needing attention, note that Example 6-4b.xls gets nearly 62 of the 100 available watts into the antenna, all while the VSWRs are 6.5 and 10.4 respectively. Clearly useful power transfer even with these high VSWRs is possible. Note, though, that the losses within the line itself are more than 38 watts under these conditions; that is more than half the amount of power absorbed by the antenna, a high number.

In practice, when a tuner is used, the tuner controls are adjusted in one of two ways: adjust for maximum reading on a convenient field strength meter, or adjust for minimum reading on the SWR meter. More needs to be said about these local SWR meters, however. The issue arises because less-astute operators may conclude that such SWR meter readings close to 1.0 implies that the transmission line standing waves have been eliminated, but this is not the case.

First, recognize that a SWR meter being used to adjust the tuner will have been connected in between the transmitter's output and the tuner's input either externally or wired that way internally. Consequently, the SWR meter is measuring the standing wave ratio that exists between the transmitter and the tuner only and is not measuring the transmission line SWR! We know that the transmission line SWR is not equal to 1.0 because we have calculated it. See Example 6-4b and figs. 6-13 and 6-14. Should readers doubt this claim, they can install a second meter in between the tuner's output and the transmission line and confirm that the SWR is what has been calculated here.

A comparison of the plots for Example 6-4a.xls and Example 6-4b.xls shows them to be of virtually the same shape. Note however, that the currents and voltages in fig. 6-13 are significantly higher than those in fig. 6-10. Don't forget that in the Example 6-4b.xls case a higher output power is being delivered and absorbed by the antenna; this obviously means the line currents and or voltage will be higher, but the line impedances remain the same. The main point is that the use of the tuner at the transmitter did not materially change the standing waves on the line. Clearly, the described system still needs attention. The matching issue at the antenna end of the line should be addressed.

The following tactics can be employed to do this and each leads to further increases in the proportion of power absorbed by the antenna: a) alter the antenna such that its input impedance more closely approximates the impedance of the transmission line; b) select a transmission line which has a characteristic impedance closer in value to the antenna impedance; c) use an impedance matching device such as a network or tuner at the input point of the antenna instead of at the transmitter's output, or d) combinations of a, b, and c. Example 6-4c employs option c.

The General Application

Data for:	Example 6-4c.xls
Input Data	
Z_0 -- T. Line Char. Imped. Ohms	50+0i
Z_L -- Rec. end load Imped.Ohms	50+0i
d -- T. Line length meters	32
α -- T. Line atten. nepers/m	0.001813
β -- T. Line ph. shift radians/m	0.34118
V_g -- Send.end source voltage	141.4+0i
Z_g -- Send.end source imped.	50+0i
Startx	0
Stopx	32
Output Results	
K_r -- Refl. Coef. at rec. end	0
abs| K_r |	0
K_s -- Refl. Coed. at send end	0
abs| K_s |	0
$VSWR_s$ -- VSWR at send. end	1
$VSWR_r$ -- VSWR at rec. end	1
Z_s -- send. end imped.	50
Y_s -- send . end Admittance	2E-002
I_s -- send. end current	1.414
abs| I_s |	1.414
I_r -- rec. end current	-0.103722830030691+1.330262080117620i
abs| I_r |	1.334299677
E_s -- send. end voltage	70.7
abs| E_s |	70.7
E_r -- rec. end voltage	-5.18614150153455+66.513104005881i
abs| E_r |	66.71498384
P_g -- Total gen/source Pwr.	199.9396
P_{sr} -- Pwr used by source	99.9698
P_s -- Pwr supplied to T. Line	99.9698
P_r -- Pwr consumed by load	89.01778136
P_L -- Pwr consumed by T. Line	10.95201864

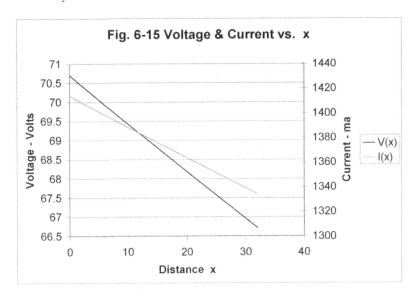

Fig. 6-15 Voltage & Current vs. x

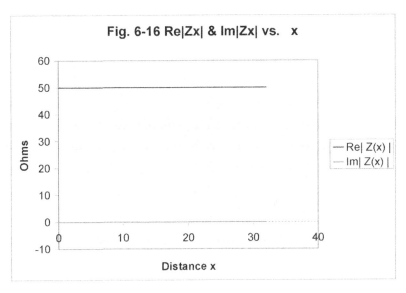

Fig. 6-16 Re|Zx| & Im|Zx| vs. x

Example 6-4c

Assume that a matching network or tuner is placed at the base of the antenna specified in Example 6-4 and adjusted such that the load seen by the cable at its antenna-end becomes $Z_L = 50 + 0i$ ohms; assume that all other specifications are identical to those in example 6-4.

Solution: The results are shown in Example 6-4c.xls and fig. 6-15 and fig. 6-16.

Example 6-4c Comments

Look first at fig. 6-16. Now that impedances are matched at both cable ends, the transmission line has an actual Z_x which equals its characteristic impedance of 50+ j0 ohms throughout its entire length; and Example 6-4c.xls shows there are no standing waves and no reflection coefficients.

There are several things to learn from fig. 6-15 as well. Because no reactances are now present, note that the product of the voltage and current at the input end of the cable, at x = 0, is 100 watts and at the antenna end it is approximately 89 watts, the same results which were computed in Example 6-4c.xls.

As cable losses have not been neglected, there should be a linear decline in voltage as x increases because of the series resistance of the cable.

At first, one might be surprised to see that the current along the line also declines linearly with increasing x. After all, doesn't all of the input current flow in the load? No! Remember that even though there are no reactive components in the line or load, two independent phenomena contribute to cable loss: the series wire resistance and the shunt conductances discussed in Chapter 3. This conductance which exists between the two transmission line conductors all along the cable's length shunts away current along the cable's entire length, meaning less and less reaches the load. Fig. 6-15 indicates this decline in current with increasing x. These declines in voltage and current are manifestations of the same slight declines in magnitude because of the finite line loss as pointed out in earlier examples.

By studying Example 6-4c.xls, it is clear that the full calculated transmitter output voltage is also allowed and even with this full output voltage, the transmitter is not over-loaded! There are no reflections; both VSWRs = 1.0 and 89% of the 100 watt output is absorbed by the antenna. Note also that the cable losses have been reduced by more than 71% from Example 6-4b.xls where the transmitter tuner is used. This matched-at-the-antenna configuration gets more power to the antenna than all of the previous instances of this problem. In fact Example 6-4c.xls is the best possible solution given this particular cable and this particular cable length. This is true because reflections have been eliminated, and should Example 6-4c.xls be recalculated with a cable attenuation α = 0, the calculations will show that 100% of the power is absorbed by the antenna.

Situation	% Pwr in ant.	db
Lossless cable matched at both ends	100	0
Normal cable matched at both ends	89	-0.506
Normal cable; Tuner used at Xmitter end	62	-2.08
Normal cable; no attempt to match	9.6	-10.2

Table 6-1   Comparing Results for Various Connections

Table 6-1 enables us to compare the results of the various ways example 6-4 has been solved. It shows the following: 1) that connecting the load without attempting any match produces poor or unacceptable results, 2) that the signal produced by using a tuner at the transmitter provides an improve-ment over a "do nothing" approach but its signal is still measurably weaker than the ideal solution; 3) that matching both ends of the cable is clearly the best solution.

As a good operator normally cannot discern the difference between two signals aurally that differ in strength by 1 decibel (db) or less, we conclude that the reduction in field strength caused by cable loss alone in the Example 6-4c.xls case would probably not be noticed in an established communication.

The conclusion is that matching transmission lines at both cable ends puts the maximum power into the load and is significantly better than using an antenna tuner at the transmitter alone. Use of the tuner at the transmitter, after all, ignores the critical system mismatch, i.e., the one at the load end of the transmission Line.

Though the reductions in load voltage and current are slight, when compared with their source levels for cables matched at both ends, as pointed out earlier, these reductions do exist because the cable attenuation α is not zero.

For the matched case, the VSWRs are equal to their ideal value of 1 at each cable end; the cable is nearly two wavelengths long, yet the line remains "flat" along its entire length. See fig. 6-15 and fig. 6-16. This means that the highest possible proportion of power is delivered to the load for this particular cable, and it does this without exceeding the specified transmitter output stage power.

The setup is not perfect, however, because 11 watts are consumed by the cable itself, but the only way to get a still higher proportion of power into the antenna is to use a cable with lower loss or to move the transmitter closer to the antenna.

Final Comments on Example 6-4

The results provided by all of the Example 6-4 s taken together indicate that a commercial broadcast station with its single output frequency would do well to spend significant time and effort reducing cable losses and tuning out any mismatches. This is because any power loss between the transmitter and the antenna results directly in diminished signal and ultimately lower station asset value to its owners. On the other hand, an amateur will most likely change frequencies many times within even a single on-air session, and with each frequency change the line/load match will also change. The amateur, then, would do well to proceed in a different way.

When creating example 6-4, it was decided that the impedance of the antenna should be significantly different from the characteristic impedance of the transmission line that fed it. By doing this, it was reasoned that the calculations would lead us into discussions of: the effects of large VSWRs and high reflection coefficients upon the line and the transmitter's output stage; show how transmitter tuners attempt to correct these kinds of problems; and finally show that by impedance matching at both transmission line ends results in the highest power transfer to the load.

Table 6-1 illustrates that for Example 6-4.xls, the practical best antenna power to be 89 watts and that a tuner located at the transmitter could get only 62 watts into the antenna, but this is with line VSWRs of 6.5 and 10. However, should the load match of example 6-4 be better with, say, VSWRs approaching 2.0 instead, a transmitter tuner could get close to 86 of the 89 watts into the antenna. (This can be easily verified by using the General application and calculating antenna power using load impedances closer to Z_o.) Thus transmitter tuners can become pretty effective when VSWRs are within a reasonable range.

Given this, an amateur would be wise to expend effort in adjusting antennas such that their input impedances produce transmission line VSWRs around 2.0 or less; but unless operation is intended for a single frequency only, trying for perfect matches or VSWRs well below 2.0 is not likely to pay off when a good transmitter tuner and low-loss cable are used.

⌘ ⌘ ⌘

7 THE REVERSE APPLICATION

In the previous chapter, it was shown that numerous quantities, and importantly Z_s, could be calculated when the transmission line parameters and the load impedance were known. However, there are occasions when the input impedance Z_s at the transmitter-end of the cable is known but the load impedance is not. In this chapter, Z_L, the load impedance, will be calculated by using the input impedance Z_s and the other cable parameters.

To do this, Equation {4-1} in Section 4 must be solved for Z_L. When this is done, the result is:

$$Z_L = Z_o \frac{(Z_o - Z_s)e^{\gamma d} - (Z_s + Z_o)e^{-\gamma d}}{-(Z_o - Z_s)e^{\gamma d} - (Z_o + Z_s)e^{-\gamma d}}$$

and the corresponding admittance is:

$$Y_L = \frac{1}{Z_L}$$

For those using calculators to arrive at solutions manually, these are the equations to use. These are also the two equations programmed into the calculation engine of the Reverse application. As with all applications when using the programs, should numerical results be desired, open the specific template file and follow the procedure outlined in chapter 5.

Example 7-1

An antenna with an unknown input impedance is connected to a 125 feet length of Belden 7810A transmission line, which has a Z_o of 50+0i ohms and a velocity factor $v_f = 0.86$. When a laboratory-grade meter, operating at 1.85 MHz, is used to measure the input impedance of the line at its transmitter end, the cable input impedance was found to be 28.43+j20.33 ohms. Assume that the attenuation factor α for 7810A cable at this frequency is 0.2 db/100 ft and find the input impedance of the antenna.

Solution: Using the methods shown in Appendix A and equation {3-4} it is determined that $d = 38.1$ meters: $\alpha = 0.0007554$ nepers/meter; and $\beta = 0.045085$ radians/meter. Next, open Reverse.xls, and click the button, enter these values into the data input form and click **Calculate**. Save the results as Example 7-1.xls. The desired unknown impedance appears as one of the outputs on the InOut1 sheet of Example 7-1.xls and is: 75.18-44.98i ohms.

The Reverse Application	
Data for:	**Example7-1.xls**
Input Data	
Z_0 -- T. Line Char. Imped. Ohms	50+0i
Z_s -- T. Line Input Z at Source end	28.43+20.33i
d -- T. Line length meters	38.1
α -- T. Line atten. nepers/m	0.0007554
β -- T. Line ph. shift radians/m	0.045085
Output Results	
Z_L -- Rec. end load Imped Ohms	75.175462990157-44.97648104345411i
abs| Z_L |	87.60270591
Y_L -- Rec. end load admitt. in S	9.79582616046156E-003+5.86071268584881E-003i
abs| Y_L |	0.011415173

Example 7-2

Jim has recently passed his general license exam and is anxious to hook up with his friends on the local 75 meter calling frequency. He has constructed a fixed single frequency, 3.855 MHz, 10 watt AM transmitter for his car to use during his commute and has it working…sort of. You have listened to his signal on the air. Its frequency is stable, the modulation is good; and there is no discernable spurious content, but his signal is weak. You invite him to drive over to see if the two of you can discover the difficulty.

When he arrives, your suspicions are confirmed: his transmitter doesn't load properly. Because Jim is a smart guy and anxious to understand what is going on with his ham gear, you proceed in a logical and teaching way toward solution of the problem.

The cable that Jim has snaked from the dash of his car, into and under the floor mats and seats, across the trunk and through the hole that leads to his rear bumper-mounted antenna is RG-8, specifically Belden 9913, but you do not know how long it is. You know that by using the techniques called out in chapter 9 that the length of the cable can be determined without having to remove it. Then Jim mentions that he purchased a 25-foot length of the cable and used the entire length, only adding a connector at the transmitter end.

After measuring the impedance at 3.855 MHz when looking into the transmission line at the transmitter end of the cable running to the antenna and finding it to be 1.426+j44.99 ohms, you ask Jim what he thought our measured value should have been.

"I think the cable is supposed to be 50 ohms. Isn't that what it should be?"

"It can only be 50 ohms *if* the cable's characteristic impedance is 50 ohms *and if* the impedance at the antenna end is matched to it. This measurement proves that your antenna is not matched to the cable."

"How can it not be matched? I used my dip meter and I am pretty sure that I adjusted my antenna to resonate at 3.855 MHz."

"You may very well have adjusted your antenna loading coil to resonate your antenna, but that is not the same thing as matching your antenna to its feed line. We know that your antenna is not matched to the line because our measurement shows that the impedance at the cable's transmitter end is not 50 ohms. Let's go inside to my computer and we will learn more about your situation."

The Reverse Application	
Data for:	**Example 7-2a.xls**
Input Data	
Z_0 -- T. Line Char. Imped. Ohms	50+0i
Z_s -- T. Line Input Z at Source end	1.426+44.99i
d -- T. Line length meters	7.62
α -- T. Line atten. nepers/m	0.0007554
β -- T. Line ph. shift radians/m	0.096184
Output Results	
Z_L -- Rec. end load Imped Ohms	0.500084678691465+3.00740928441497E-004i
abs$\mid Z_L \mid$	0.500084769
Y_L -- Rec. end load admitt. in S	1.99966061939518-1.20255592076604E-003i
abs$\mid Y_L \mid$	1.999660981

Using equation {3-4} and the techniques described in Appendix A, you find α to be 0.0007554 nepers/meter and β to be 0.096184 radians/ meter; then you calculate d to be 7.62 meters. You enter these numbers and Z_0 into a copy of Reverse.xls. After clicking **Calculate**, you save the results as Example 7-2a.xls. As seen on Example 7-2a.xls the impedance was found to be 0.5 +j0 ohms.

You ask, "Jim, from the numbers we have just gotten from the Reverse application, can you determine if your antenna is adjusted to resonance at 3.855 MHz?"

"I think there should be no reactance when the antenna is resonant."

"That's correct, Jim! And sure enough the reactance term of the antenna impedance we just found is indeed essentially equal to zero. Good." You continue, "Your antenna is resonant but we have another problem. The real part of your antenna impedance needs to be 50 ohms, i.e. equal to the characteristic impedance of your transmission line, and not the 0.5 ohms that it actually is. That is why your transmitter can't load much power into the line. Let me show you something."

The General Application

Data for:	Example 7-2b.xls	
Input Data		
Z_o -- T. Line Char. Imped. Ohms	50+0i	
Z_L -- Rec. end load Imped.Ohms	.5+0i	
d -- T. Line length meters		7.62
α -- T. Line atten. nepers/m		0.0007554
β -- T. Line ph. shift radians/m		0.096184
V_g -- Send.end source voltage	44.72+0i	
Z_g -- Send.end source imped.	50+0i	
Startx		0
Stopx		7.62
Output Results		
K_r -- Refl. Coef. at rec. end	-0.98019801980198	
abs| K_r |		0.98019802
K_s -- Refl. Coed. at send end	-0.101509788126647+0.963646664717676i	
abs| K_s |		0.968978396
$VSWR_s$ -- VSWR at send. end		63.47119841
$VSWR_r$ -- VSWR at rec. end		100
Z_s -- send. end imped.	1.42583138050778+44.9894603065136i	
Y_s -- send . end Admittance	7.03737359781177E-004-2.22051249866661E-002i	
I_s -- send. end current	0.492595177250237-0.430942788461745i	
abs| I_s |		0.654493465
I_r -- rec. end current	0.654379141247582-0.589067947127734i	
abs| I_r |		0.880461871
E_s -- send. end voltage	20.0902411374881+21.5471394230873i	
abs| E_s |		29.46009176
E_r -- rec. end voltage	0.327189570623791-0.294533973563867i	
abs| E_r |		0.440230936
P_g -- Total gen/source Pwr.		38.88859638
P_{sr} -- Pwr used by source		21.41808478
P_s -- Pwr supplied to T. Line		0.610771548
P_r -- Pwr consumed by load		0.387606553
P_L -- Pwr consumed by T. Line		0.223164994

As in Example 6-4.xls you calculate the input voltage when 10 watts are inputted into the line and found it to be 22.36 volts, making the transmitter signal source voltage to be 44.72+0i volts, You enter this and the antenna impedance of $Z_L = 0.5 + j0$ ohms, which we had just found from Example 7-2a.xls, into a copy of General.xls, along with the same d, β, and α that we used in Example 7-2a.xls. You click **Calculate** and save this new file as Example 7-2b.xls.

The General Application

Data for:	Example 7-2c.xls
Input Data	
Z_o -- T. Line Char. Imped. Ohms	50+0i
Z_L -- Rec. end load Imped.Ohms	50+0i
d -- T. Line length meters	7.62
α -- T. Line atten. nepers/m	0.0007554
β -- T. Line ph. shift radians/m	0.096184
V_g -- Send.end source voltage	44.72+0i
Z_g -- Send.end source imped.	50+0i
Startx	0
Stopx	7.62
Output Results	
K_r -- Refl. Coef. at rec. end	0
abs$\mid K_r \mid$	0
K_s -- Refl. Coed. at send end	0
abs$\mid K_s \mid$	0
$VSWR_s$ -- VSWR at send. end	1
$VSWR_r$ -- VSWR at rec. end	1
Z_s -- send. end imped.	50
Y_s -- send . end Admittance	2E-002
I_s -- send. end current	0.4472
abs$\mid I_s \mid$	0.4472
I_r -- rec. end current	0.330461466330029-0.2974793132995506i
abs$\mid I_r \mid$	0.444633245
E_s -- send. end voltage	22.36
abs$\mid E_s \mid$	22.36
E_r -- rec. end voltage	16.5230733165014-14.8739656649753i
abs$\mid E_r \mid$	22.23166225
P_g -- Total gen/source Pwr.	19.998784
P_{sr} -- Pwr used by source	9.999392
P_s -- Pwr supplied to T. Line	9.999392
P_r -- Pwr consumed by load	9.884936129
P_L -- Pwr consumed by T. Line	0.114455871

"Look at these numbers Jim. Your reflection coefficients are between 97% and 98%. (See Example 7-2b.xls) These high values are preventing power from entering your transmission line and your antenna. You are getting only about 0.6 of a watt into your transmission line at the transmitter and only 0.39 of a watt into the antenna itself. All of the energy which you are trying to radiate is being dissipated in your transmission line and in the output stage of your transmitter. It is almost certain that your low transmit duty cycle combined with having a vacuum tube output stage in your transmitter, is the reason you have not burned up your output stage before now. It is dissipating power at nearly twice it's rating! We must improve the match of your antenna to the line so that the line will begin to accept the power from your transmitter and move it to the antenna."

You suggest that he use an L-C network at the base of his antenna to improve the match between his antenna and transmission line. By using the techniques shown in Appendix B you calculate values for L and C and find them to be 0.2054 microhenries and 8.216 nanofarads, respectively.

After connecting these components as close to the antenna base as possible in the way described in Appendix B and using a SWR meter to tune them, the SWR is reduced to a nominal value of 1 at the antenna. After this had been completed, the transmitter loaded up as it should.

As the match has been achieved, you change Z_L to $50+0i$ in Example 7-2b.xls, click **Calculate,** and save the result under the new name of Example 7-2c.xls.

A comparison of the results obtained for the first or unmatched line connection at the antenna, shown by Example 7-2b.xls, with the results for a matched antenna connection shown

Watts	Antenna & line not Matched	Antenna Matched to Line
Power into Line	0.611	10
Power into Antenna	0.388	9.88
Line Loss Power	0.223	0.114

Table 7-1   Power Comparisons

in Example 7-2c.xls is displayed in table 7-1. The ratio of power in the antenna with a match to power in the antenna without a match is more that 25. This means that Jim has improved his signal strength by more than 25 times, an equivalent gain of 14 db!

Comments on Examples 7-2a, 7-2b, 7-2c and 7-2d

The calculations show a decided improvement when the antenna is matched to its feed line, but what about this matching network? Is it certain that such a network can work such magic?

According to Example 7-2c.xls, this network eliminated line reflections, made the VSWRs = 1.0, and now causes transfer of nearly all of the transmitter power to the antenna.

To better understand the matching device, let us look more closely at the calculations used to find the values for L and C. To do that, a new spreadsheet called Example 7-2d.xls was created; it contains a number of calculations relating to the L-C matching network. (A copy of Example 7-2d.xls is included in Appendix C and an Excel version with all equations is included on the CD.)

The first entries in the Example 7-2d.xls spreadsheet use equations {B-6} and {B-7} from Appendix B to calculate L and C; the results are displayed in the upper-left corner of the sheet (Cells B5 through C11). Because equation {B-1} in Appendix B is the expression for the input impedance of the combination of the antenna and the matching network, and because all the values for the quantities in that equation are known, it can be used to find numerical values for the input impedance of that combination. By so doing, note that the calculated results verify that the input impedance of the combination equals the transmission line impedance. (see cells E9 and E10 of Example 7-2d.xls) Stated another way, these calculations prove that the matching network works. That is, the real part of the calculated impedance made up of the combination of the antenna impedance and the matching network impedance exactly equals the characteristic impedance of the line, and the imaginary part of this impedance combination is very small (less than 1.8×10^{13} ohms or essentially zero).

In the real world, however, practical real values rarely precisely equal calculated values. Therefore it is reasonable to study how much variability in L and C in the network can be tolerated before the antenna/matching network performance becomes unacceptable.

The first part of this analysis checks the bandwidth of the antenna/matching network combination when the antenna used is the one described in Example 7-2a.xls. As stated earlier, because Appendix B equation {B-1} represents the impedance seen looking into the matching network/antenna combination, with frequency as one of the variables, we will use the previously calculated values for C, R and L in equation {B-1} and produce a chart of impedance values vs. frequency.

Check Bandwidth of Match Network

L=	2.054E-07	
C=	8.216E-09	
freq	VSWR	Pwr in Ant
3.845E+06	1.053	9.878
3.847E+06	1.042	9.881
3.849E+06	1.031	9.883
3.851E+06	1.021	9.884
3.853E+06	1.010	9.885
3.855E+06	1.000	9.885
3.857E+06	1.010	9.885
3.859E+06	1.021	9.884
3.861E+06	1.031	9.883
3.863E+06	1.042	9.881
3.865E+06	1.053	9.878
3.867E+06	1.064	9.876

Check Effect of Inductor Tolerance

f=	3.855E+06		
C=	8.216E-09		
%	L	VSWR	Pwr in Ant
90	1.849E-07	2.606	7.924
93	1.900E-07	2.075	8.677
95	1.951E-07	1.636	9.309
98	2.003E-07	1.282	9.734
100	2.054E-07	1.000	9.885
103	2.105E-07	1.282	9.734
105	2.157E-07	1.636	9.309
108	2.208E-07	2.075	8.677
110	2.259E-07	2.606	7.924

Check Effect of Capacitor Tolerance

f=	3.855E+06		
L=	2.054E-07		
%	C	VSWR	Pwr in Ant
90	7.394E-09	1.990	8.801
93	7.600E-09	1.559	9.413
95	7.805E-09	1.250	9.763
98	8.010E-09	1.064	9.875
100	8.216E-09	1.000	9.885
103	8.421E-09	1.059	9.877
105	8.626E-09	1.240	9.771
108	8.832E-09	1.544	9.433
110	9.037E-09	1.970	8.830

Table 7-2. Summary of Matching Network Performance

By studying {B-1} we note that both of its real and imaginary terms have the same denominator. Therefore, Example 7-2d.xls's spreadsheet calculation formulas can use the variables appearing in Cells C5 through C11 and compute the denominator which is labeled "denom" in Cells D15 through D26.

In the same way "Re(num)" and "Im(num)" can be calculated. Next Re(Zi) and Im(Zi) are calculated by dividing "Re(num)" and "Im(num)" by "denom" yielding Re(Zi) and Im(Zi). These two quantities are combined into Zi in column J. All these calculations are done for each row in the example. These data are shown in Cells D15 through J26.

Each calculated impedances is used as a load impedance Z_L in a copy of Example 7-2c.xls, without changing any of its other parameters. This produces values of load VSWR and antenna power for each impedance. Thus we now have a chart of VSWR and antenna power vs. frequency for the load combination. This is shown in cells L15 through M26 of Example 7-2d.xls. These values have also been copied to the "check bandwidth" portion of table 7-2.

Continuing the analysis, fix the frequency at the operating value of 3.855 MHz, use the ideal value of C; let L vary from 90% to 110% of its ideal value, and then calculate the input impedance of the combination for each L. Again use a copy of Example 7-2c.xls to compute values of VSWR and antenna power for each value of L. (see Cells L33 through M41). These results are also shown in the "check inductor tolerance" section of table 7-2.

Finally, repeat this process with L fixed and the frequency fixed at 3.855 MHz while C is allowed to vary from 90% to 110% of its original calculated value. These results constitute the third data set shown on the Example 7-2d.xls sheet. See cells L46 through M54 and the "check capacitor tolerance" portion of table 7-2.

Now look at the results. The "check bandwidth" portion of table 7-2 where L and C are held fixed shows hardly a bump in the VSWR or antenna power while varying the frequency over a 22 KHz range. This shows that the matching network has good bandwidth for this particular antenna under these conditions. However, should the operating frequency be changed by this amount, Jim's mobile antenna loading coil would almost certainly require readjustment to maintain resonance and to keep the antenna's impedance even close to the 0.5 +j0 ohms, upon which the matching network calculations are based.

Variations in L and C around their ideal values do not fare as well. The data show that an actual value of L which differs from its ideal calculated value by only 7% will raise the line VSWR above 2.0, and a 10% error in C will cause a similar line VSWR. These results suggest that care and precision are required in setting the values for L and C for this matching circuit and its 75 meter mobile antenna.

⌘　⌘　⌘

8 THE UNKNOWNCABLE APPLICATION

This Book is about solving transmission line problems and as chapters 3 and 4 have pointed out, the transmission line itself plays an integral part in the behavior of the circuits that use them. This means that to begin solving any problems, the parameters that define a transmission line must be known. This chapter describes several ways in which transmission line parameters can be determined.

Parameter Estimation

These first two methods are the least accurate of those discussed. In the simplest method, the user selects the cable listed in a reference similar to the tables in the *ARRL Antenna Book* (ARRL 2005: 24-18, 24-19), which has characteristics most like those of the unknown cable, copies the appropriate parameters from the printed table, measures the cable's length, and declares that the needed parameters have been found.

The second method may provide more accuracy or at least a confirmation for the value of Z_o obtained in the first method. It uses physical dimension measurements and an estimate for the value of relative permittivity that applies to the cable in question. Once these values are obtained they are used in one or the other of two equations from (Kraus 1953:428–429) depending upon the type of cable involved.

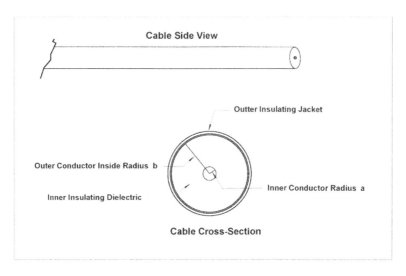

Figure 8-1 Coaxial Cable Dimensions

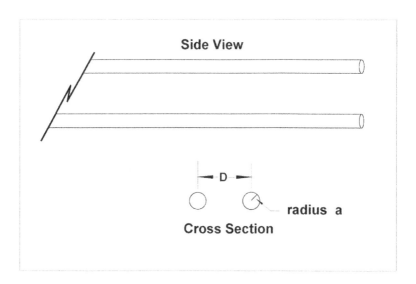

Figure 8-2 Two-Wire cable Dimensions

The measurements yield values for b and a for the coaxial cable case and D and a for the two-wire cable case. See figs. 8-1 and 8-2.

The relative permittivity is obtained by estimating the velocity factor v_f that applies to the unknown cable. Again, the tables in the *ARRL*

Antenna Book are a good source for values of v_f. To find an estimate for v_f, try to indentify the cable's dielectric material by studying the material types listed in the table, and note the velocity factor v_f corresponding to that material type.

Next, note that from equation {3-2} and the discussion nearby that the following relation can be written:

$$v_f = \frac{1}{\sqrt{\mu_r \, \varepsilon_r}}$$

Recognizing that for normal cables $\mu_r = 1.0$ we get

$$\varepsilon_r = \frac{1}{\left(v_f\right)^2}$$

Finally, use the measured dimensions b and a and our newly found value for ε_r in the following equation for coaxial cables to find Z_o:

$$Z_o = \frac{138}{\sqrt{\varepsilon_r}} \log \frac{b}{a}$$

and the measured dimensions D and a for the two-wire cable in the following equation and find Z_o:

$$Z_o = \frac{276}{\sqrt{\varepsilon_r}} \log \frac{D}{a}$$

The accuracy of the value obtained for Z_o by these methods will be limited by the normal variations in the actual dimensions for any real cable, measurement errors, and the uncertainty of the velocity factor selected for use in these equations, but the results should provide an estimate for Z_o that is within 15% or so of its real value.

Concerning the remaining four parameters: the length d can be measured with a tape measure; α, can be estimated by study of the transmission line tables from which v_f was previously selected; and finally, given the operating frequency and the estimate for v_f, β can be calculated by using equation {3-3}. While significantly better results for these parameters can be found by the method described next, circuit calculations using the parameters obtained by these estimates can provide at least rough-cut results.

Obtaining Parameters via Electrical Measurements

This method can determine all of the key transmission line parameters. It relies upon electrical and length measurements of the unknown cable. (see Chipman 1968, 135) In this scheme an impedance-measuring instrument or bridge is used. The bridge must be of the balanced type when working with a two-wire transmission line but of the unbalanced type when measuring a coaxial cable. The cable can be of any length providing it is long enough to produce measurable quantities for the bridge at the frequency being used. Cables longer than a few wavelengths, can however, be difficult to characterize using this method. See the discussion accompanying the following examples.

To use this method, perform the following tasks.

1. Measure the cable's physical length d and express it in meters.
2. Select a frequency such that using equation {3-4} makes the signal wavelength within the cable about a wavelength long or less (but do not select a frequency that makes the cable close to a length of $n\lambda/4$ where n is odd; see chapter 9), and keep that frequency constant for the following steps.
3. Measure the input impedance at end #1 of the cable with end #2 short-circuited and record Z_{sc}.
4. Measure the input impedance at end #1 of the cable with end #2 open circuited and record Z_{oc}.

Using d and the results from the two impedance measurements the following quantities can be calculated by:

$$Z_o = \sqrt{(Z_{sc})(Z_{oc})}$$

which is an impedance expressed in ohms.

$$\alpha = \frac{1}{2d} \ln\left[\frac{1 + \sqrt{Z_{sc}/Z_{oc}}}{1 - \sqrt{Z_{sc}/Z_{oc}}}\right] \qquad \{8-1\}$$

is the attenuation in nepers/meter and

$$\beta = \frac{1}{2d}\left[phase\ angle\ of \left(\frac{1 + \sqrt{Z_{sc}/Z_{oc}}}{1 - \sqrt{Z_{sc}/Z_{oc}}} \right) + 2\pi n \right] \qquad \{8-2\}$$

is the phase shift in radians per meter.

Finding values for Z_o and α is straight forward: use a calculator with vector function capability or a spreadsheet to carry out the calculations indicated or use the application described below. It is more difficult to find β because there is no direct mathematical basis for determining n in equation $\{8-2\}$.

Study of this situation indicates that n is unknown because the cable's velocity factor v_f is also not known. Consequently, additional steps are required to find v_f then n and eventually β. Conceptually, the method is to try various values for v_f and n in equation $\{8-2\}$ until α, β, v_f and n, taken all together, force the equations to fit the measured values for Z_{sc} and Z_{oc}. However, the following procedure provides a straightforward way to obtain results logically.

To begin, it is important to recognize that all values for β must have the units of radians/meter not degrees/meter. Calculate a starting value for β. To do this, use equation $\{3-3\}$ while assuming $v_f = 1.0$. By doing this, we are assuming that the transmission line is operating as if it exists in free space, i.e. $\beta_{free-space} = (2\pi f)/c$. As we proceed, the actual beta for the cable in question will be found.

Next, make a first guess or estimate for an actual v_f. The best way to do this is to study the transmission line tables that show cable data, including v_f for the various cable types with characteristics similar to the unknown cable being measured. The type of dielectric used in the cables is important in these comparisons. Considering these factors, make a first estimate for v_f.

Now with starting values for beta and the velocity factor in hand, use them to compute a new value for beta, call it $\beta t = \beta_{free-space}/v_f$. This value for βt will then be used to choose the final β value from a list of betas which will now be calculated in a different way.

Note that the expression for β in equation $\{8-2\}$ has two terms, which together can be written as follows:

$$\beta = \frac{ATAN\frac{X}{R} + 2\pi n}{2d} \qquad \{8-3\}$$

where X is the imaginary or reactive part of the "phase angle portion" of β and R is the real part of the "phase angle portion". Using equation {8-3}, generate a table of n vs. beta, one beta for each value of n where $0 <= n <= {\sim}20$.

Important: Remember that ATAN (arc tangent with complex arguments) can produce values in all four trigonometric quadrants and care must be taken to value ATAN according to the quadrant into which it falls for all cases.

From this list of ~21 beta values, look for that beta which is closest in magnitude to βt and select it. This selected value is the unknown β we seek, i.e., this is the β that characterizes the phase factor for the unknown cable. The correct value for n is that value that is associated with the β just selected from the list. Finally, compute the v_f for the unknown cable by using the value for β just selected in equation {3-3} above, i.e., $v_f = (2 \pi f)/\beta c)$.

As just described, once Z_{sc} and Z_{oc} have been measured and the cable length determined, α can be calculated directly using equation {8-1}. But as we have seen, getting to β requires, in general, repeated calculations involving the complex quantities in equation {8-3}. However, the preprogrammed UnknownCable application will easily calculate Z_o, α, β, v_f and n directly, without any repetitive data entry. (This application is included on the recommended CD.) See chapter 5 for additional information on using the preprogrammed applications.

Example CableShort

At a flea market Joe learns that he could purchase a coil of coaxial cable for about half the normal price of similar cable. As offered, it had two problems. It is unmarked without any information concerning its characteristics; and second, the cable had been cut almost into two pieces at about 6 feet from one end; but otherwise it appeared to be in excellent condition. Joe notices that the cable and conductor sizes are similar to those he frequently uses, and the dielectric also looks the same. He decides to purchase the coil of cable and attempt to characterize it himself.

The UnknownCable Application	
Data for:	**ExampleCableShort.xls**
Input Data	
Frequency -- Hz	1.40E+07
d test cable len. -- m	1.8
Z_{sc} -- ohms	.2443+35.25i
Z_{oc} -- ohms	.4914-70.91i
Estimated v_f	0.7
Output Results	
Z_0 -- ohms	49.99697539872-1.46308850518486E-005i
α -- nepers/meter	0.001813183
βt -- radians/meter	0.419169645
n -- indx of selec'd β	0
Calculated β	0.34117
Calculated v_f	0.86003

Solution: Once back at his shack, Joe severs the cable where it has been damaged. He squares each end of the short piece of cable, measures its length and finds it to be 1.8 meters. He sets the frequency of his antenna analyzer bridge to 14 MHz and measures the Z_{sc} = 0.2443+35.25i ohms and Z_{oc} = 0.4914-70.91i ohms. Next, he opens the UnknownCable.xls file into an open Excel environment and clicks **Enter/Change Data.** This causes the UnknownCable data input form to appear. He then types in his measurement data. As described earlier, an estimated value for v_f is required, and for this, he enters 0.7. When he had completed this, he clicks **Calculate** on the form and saves his results as ExampleCableShort.xls. (Open this file and click its button to see how Joe entered his data.) The results obtained appear in ExampleCableShort.xls. As can be seen, the application produces values for Z_o, α , *n*, βt, β and a newer, more accurate estimate for v_f as well.

For good measure, Joe decides to use that more-accurate value for v_f which was calculated by the application. He enters that calculated value for v_f = 0.860 in place of his earlier estimate into the input form as his new estimated value for v_f and clicks **Calculate** again. The only value which changes when he did this was βt, the test beta; and it became equal to the final calculated value for β. All of the other cable parameters remained as they were.

The program appears to have computed the correct cable characteristics, but Joe wonders what would happen if he had chosen a beginning estimate for v_f that was significantly different from 0.7. He tried using estimated values for v_f ranging from 1.0 down to 0.3, clicking **Calculate** for each value entered. The application produced the same cable parameters for each v_f he tried and always suggested the same better estimate for $v_f = 0.860$. He reasoned, v_f can never be as high as 1.0 because that would make the wave velocity on the cable equal to the velocity of a wave in free space; and from his studies of the transmission lines tables, he knew that v_f could also not be as low as 0.3 because virtually no cables of any type listed in the tables have v_fs as low as that. Thus Joe felt confident that he now knew the parameters of his cable, but he decided to perform one more test.

Example CableLong

Joe squares off both ends of the longer part of his cable, measures it, and finds it to be 30 meters long. He then measures the impedances and finds $Z_{sc} = 5.704+52.26i$ and $Z_{oc} = 5.159-47.27i$ ohms. Next he opens UnknownCable.xls, clicks the button, and enters these data. Because the short cable had found the v_f to be 0.86, he entered this as his first estimate, clicks **Calculate** and saves the file as ExampleCableLong.xls (which shows his results).

Joe is pleased because his actions produce the same cable parameters for the short cable *and* for the long cable! (See ExampleCableLong.xls.) For completeness, he decides to see how sensitive the long cable results are to values of estimated v_f. He starts with an estimated $v_f = 1.0$ and reduces it a step at a time, recalculating for each value used. This time the results were the same for $v_f=1.0$ down to $v_f = 0.75$; but below 0.75 the application file suggested a calculated value for v_f which was even lower than this already low estimate. Joe recognized that this outcome indicated that the results for these inputs, were suggesting that the estimated values of v_f below 0.75 are out of range. Thus under these conditions an estimate for v_f larger than 0.75 but less than 1.0 is required.

Joe realizes that the long cable at 14 MHz is more than 1.6 wavelengths long; and that this means that the measurements had to have been affected by the increased cable losses. It seemed reasonable, therefore, that the range of estimated v_fs that yields correct results would be more narrow for the long cable.

In the end Joe was satisfied that he had learned something, found the unknown cable characteristics, and saved some money. His cable possessed the following characteristics at 14 MHz:

Z_o = 50 +0i
α = 0.001813 nepers/meter
β = 0.3412 radians/meter
v_f = 0.86

The UnknownCable Application	
Data for:	**ExampleCableLong.xls**
Input Data	
Frequency -- Hz	1.40E+07
d test cable len. -- m	30
Z_{sc} -- ohms	5.704+52.26i
Z_{oc} -- ohms	5.159-47.27i
Estimated v_f	0.86
Output Results	
Z_0 -- ohms	49.9975713013655-1.87409103205027E-004i
a -- nepers/meter	0.001813153
βt -- radians/meter	0.341184595
n -- indx of selec'd β	3
Calculated β	0.34118
Calculated v_f	0.86001

An additional mechanism could have been at work for the longer cable. As the reader discovered in chapter 6, impedance levels increase without limit as line lengths approach n (¼) wavelengths for n = 1, 3, 5... etc. This effect will also be obvious from a glance at a Smith Chart. Consequently, if an attempt is made to measure either of the two impedances required for this method when the cable's length approaches one of these wavelengths, the measured results will likely be swamped by the very large cable impedance values. Should the real or imaginary impedances measured for Z_{sc} or Z_{oc} exceed 200 to 300 ohms when measuring an ordinary coax line, this is a strong indication that the measurement is being made too close to one of these undesirable wavelengths. If this occurs, shift the measurement frequency, until the impedance values are more reasonable, and proceed through the steps again.

The scheme just described, to characterize unknown cables by making electrical measurements, can work well and provide good accuracy, but as the real examples presented here demonstrate, a good estimate for v_f, especially for longer cables, is important. It is also true that the longer the cable, the narrower the range of estimated v_f which will yield the correct result. In addition, the impedance measuring device may need to be capable of making accurate low-value measurements. In all cases the accuracy of the measuring equipment and the care exercised when performing the measurements directly determine the accuracy of this type of cable characterization.

Manufacturer's Data

The preferable method of obtaining the transmission line parameters is to use data supplied by the transmission line's manufacturer. However, some care is required because of the differing units that suppliers may make available.

Perhaps the most troublesome characteristic is the Cable attenuation α. This book uses attenuation α specified in nepers/meter because this unit came out of Heavyside's transmission line solution. However, some manufacturers and table publishers will use other units for α.

For example, the *ARRL Antenna Book* (2005) lists cable attenuations in db per 100 ft for α, implying that some conversion factor is required. To make such a conversion, select its table value fitting the problem of interest (say 0.4 db per 100 feet), and multiply that selection by a proper conversion factor.

The factor which converts attenuation expressed in db per 100 ft into nepers per meter is derived in Appendix A and is 0.003777206517, therefore:

$$\alpha = \left(0.0037772065\ 17 \frac{neper\ ft}{db\ meter} \right)\left(0.4_{db/100ft} \right)$$

$$= 0.0015108826 \frac{nepers}{meter}$$

The phase shift factor β and wavelength λ are determined by using the v_f provided by the manufacturer in equations {3-3} and {3-4} in chapter 3.`

The use of conversion factors may be required for any of the transmission line characteristics. And in some cases, conversion factors may need to be derived. See Appendix A.

⌘   ⌘   ⌘

9 THE LENGTH APPLICATION

Methods for finding the cable parameters Z_o, α, β and v_f were described in the previous chapter but one critical parameter is yet to be discussed: the cable's length. In many (perhaps even most) cases, one needs to stretch out the cable and measure its length using a tape measure. Invariably, however, there will be the case when applying a tape measure is not a viable option to determine length. What to do?

The answer lies in measuring the input impedance at one end of the cable with its second end short circuited. By substituting this measured impedance and the other known parameters, α, β and v_f for the line in question, into the Transmission Line Equation solution, the cable's length can be calculated.

To do that, start with the Transmission Line Equation as written in equation {4-1} and express it in a different form. That is, enter $0+0i$ for each Z_L in {4-1} and divide both sides by Z_o. Next, change Z_s, the sending end impedance to the short circuit impedance Z_{sc} that was obtained by measurement. This leaves:

$$\frac{Z_{sc}}{Z_o} = \frac{e^{\gamma d} - e^{-\gamma d}}{e^{\gamma d} + e^{\gamma d}} = TANH(\gamma d)$$

From this last relationship two expressions can be written, one for the real part of $TANH(\gamma d)$, which is:

$$Re\left[\frac{Z_{sc}}{Z_o}\right] = \frac{SINH(2\alpha d)}{COSH(2\alpha d) + COS(2\beta d)} \qquad \{9-1\}$$

and the second for the imaginary part of $TANH(\gamma d)$:

$$IM\left[\frac{Z_{sc}}{Z_o}\right] = \frac{SIN(2\beta d)}{COSH(2\alpha d) + COS(2\beta d)} \qquad \{9-2\}$$

Theoretically, either {9-1} or {9-2} will yield the same result for the length d. As a practical matter, however, the imaginary expression in equation {9-2} is likely to have more meaningful numerical magnitudes and thus provide more accurate results; therefore {9-2} is selected as the equation for use.

As stated above, Z_o, α, and β are assumed to be known quantities; and Z_{sc} is obtained by measurement, leaving d as the only unknown quantity to be determined using {9-2}. Note however, that solution of even this simplified version of equation {4-1} is difficult because no simple algebraic expression for d can be written from equation {9-2}. In other words, the expression is said to be transcendental (i.e., the solution requires a trial and error approach). A trial-and-error method can be carried out by using a vector function calculator, with patience; but literally dozens of trials may be required. This suggests the need for a computer application where the length can be found directly without manual repetitive data entry. The algorithm developed for this purpose is the Length application, which is straightforward and easy to use, but it has issues.

As can be seen from any Smith Chart, the real and imaginary impedance values, near the 0.25 wave-length point and odd multiples of a 0.25 wavelength for any cable can increase without limit. This means that for a given measurement frequency, there are cable lengths centered around these 0.25 wavelength points in which large impedances will interfere with and possibly invalidate the results of any measurement algorithm. This means that measurements at these wavelengths must be avoided.

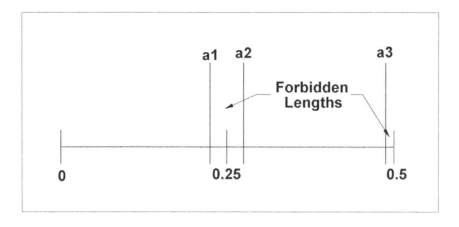

Figure 9-1 Acceptable Cable Wavelengths

Because of this the Length application algorithm was purposely designed to work only with cable lengths less than 0.5 wavelengths long; and it will also not work within a band of lengths from just short of a 0.25 wavelength up to just longer than a 0.25 wavelength. See fig. 9-1. Thus the wavelengths between a1 and a2 in fig. 9-1 and those between a3 and 0.5 must be avoided. What are the values for a1, a2, and a3?

Actually, the wavelengths between a3 and 0.5 in fig. 9-1 are not forbidden. It is just that the measurement frequency must not be set to allow the cable wavelength to reach 0.5 or any value larger than 0.5. As the measurement wavelength approaches 0.5, the magnitudes of the measured impedance values may become too small for meaningful measurement anyway, this region should be avoided.

The values for a1 and a2 do, however, mark the boundaries for wavelengths which are not allowed to be used with the Length application and their values depend on the cable length d and the cable attenuation factor α. Using a good quality cable with α values of 0.00226 nepers/meter (0.6 db/100 feet) or smaller and $d \sim= 30$ meters or less, a1 is 0.23 and a2 is 0.27; for cables with 5x larger α, a1 is 0.19 and a2 is 0.31. On the other hand, should the cable be shorter, say$\sim= 3$ meters with the larger α, a1 and a2 will again be $\sim=0.23$, and 0.27 respectively.

By now, the reader may be perplexed. Doesn't that 0.5 maximum wavelength severely limit the usefulness of the Length application? And isn't keeping up with a1, a2, and a3 a real hassle? Hold on. Provided the measurement bridge can be used at a frequency as low as, say, 1 MHz, the length of cables approximating 125 meters (410 feet) can be determined using this technique. Should the bridge be capable of measuring at 400 kilohertz (kHz), then lengths of cables 200 meters (656 ft) long can be determined. How many amateur installations will need to determine lengths exceeding these numbers? When using the Length application, one need not keep up with a1, a2, or a3 at all. The procedure for using the application takes care of that matter. The concept of the "forbidden" wavelength boundaries is presented only to make the reader aware of the issue.

All this means that the application must control conditions such that forbidden wavelengths will not be used for impedance measurements. This implies that the measurement frequency must be selected such that the unknown cable length will be safely away from the 0.25 wavelength and 0.5 wavelength points when the impedance measurement is made.

Numerous trials have demonstrated that the effects of high α and longer cable lengths affect the measurements least when impedance measurements are made at about 0.135 and or 0.39 wavelengths. Therefore, unless one needs to move closer to the 0.25 wavelength or to the 0.5 wavelength points for whatever reason, the measurement frequency should be set so as to make the unknown cable length approximate one or the other of these two values of wavelength. It is not necessary, however, to

	When using Eqn {9-3}	When using Eqn {9-4}
Unknown cable length is:	1 to 20 meters	5 to 200 meters
Adjust freq. until Im[Zsc] is:	$\sim 0.8\ Zo \leq$ Im[Zsc] $\leq$ $\sim 2\ Zo$	$\sim -2\ Zo \leq$ Im[Zsc] $\leq$ $\sim -0.04\ Zo$

Table 9-1 Finding the Measurement Frequency

achieve these wavelengths exactly. In fact, the Length application will work correctly when the actual wavelengths are anywhere between 80% and 120% of either 0.135 and or 0.39.

Thus, to use the Length application, first make an estimate for the unknown cable length. Select the column in table 9-1 that applies to that length estimate, and use the equation listed at the top of the selected column to compute the first-cut measurement frequency, i.e., use either equation {9-3} or equation {9-4}:

$$f = \frac{(0.135)v_f c}{est.length} \qquad \{9-3\}$$

$$f = \frac{(0.39)v_f c}{est.length} \qquad \{9-4\}$$

With the measurement frequency now known, execute the following series of steps:

- Short circuit one end of the subject cable.
- Measure the input impedance at the open end using the measurement frequency just determined.
- Check table 9-1 again, this time use the measured results in order to determine if adjustment of the measurement frequency is required. The imaginary part of the impedance measured must be within the range indicated. Should an adjustment be required, remember that $Im[Z_{sc}]$ increases with increasing frequency.
- When $Im[Z_{sc}]$ is within the proper range as indicated by table 9-1, record this new measurement frequency.
- Using equation $\{3-3\}$ and the final measurement frequency, calculate β.
- Measure and record the short circuit impedance Z_{sc} at this frequency.

The same impedance bridge rules listed in Chapter 8 must be followed here as well. Remember the accuracy and precision of the length obtained by these procedures is determined directly by the precision and accuracy of the impedance bridge measurements and the accuracy, precision and stability of the measurement source signal, implying that the measurement should be approached with care. Generally, four to five digits of frequency and impedance accuracy will be required to produce a length precision of four digits.

Having obtained Z_o, Z_{sc}, α, and β, start Excel, open Length.xls, click on **Enter/Change Data,** and type these values into the Length input data form. Click **Calculate** on the data input form. This will cause the desired results to be displayed on the Length.xls InOut1 sheet. (See chapter 5 for specific instructions on the use of the computer applications.)

Example Length1

Ted installed a 20 meter vertical antenna some one hundred feet to the southwest of his shack about a year ago. He fed this antenna with RG-8U Type PE cable that he buried underground. After a year's operation Ted has concluded that his output power is noticeably less on the low end of the band and has decided to characterize the behavior of his antenna across the band before making any changes. To do this, he decides to use the application described in example 6-4 of chapter 6 to analyze his situation. He realizes that he doesn't know one of the required critical parameters: the length of his transmission line! Because the cable is expensive, has been performing well, and might be damaged should it be dug up to measure its length, he applies the Length.xls application to obtain its length.

From the spec sheet for his RG-8U Type PE cable he records the following specifications:

Z_o = 52 ohms
v_f = 0.84
α = 0.7 db/100 ft or 0.002644 nepers/meter (from equation {A-12});

Using 100 feet as his estimated length, Ted converts the length into meters, i.e., his estimated length = 100ft = 30.48 meters. (See Appendix A for conversion methods.) The estimated length indicates that he should use equation {9-4}, so he enters his length estimate and finds f to be 3.2222 MHz. He then measures the shorted cable input impedance at this frequency and obtaines: Z_{sc} = 11.9-77.03i ohms. He notes that Im[Z_{sc}] fits within the range required by table 9-1 and thus makes no adjustment in the measurement frequency.

Ted computes β at the measurement frequency using equation {3-3} and finds it to be 0.080396 radians/meter. He opens Length.xls and clicks the button. After entering his data into the Length data input form and clicking **Calculate**, he saves his results as Length1.xls.

Length Transmission Line Application		
Data for:	**Length1.xls**	
Input Data		
Z_0 -- T. Line Char. Imped. Ohms	52+0i	
Z_{sc} -- Send End Imped. Ohms	11.9-77.03i	
α -- T. Line atten. nepers/m		0.0026440
β -- T. Line ph. shift radians/m		0.0803960
Output Results		
Length meters		26.8301
Length in wavelengths		0.343303

Note that Length1.xls shows the length to be 26.83 meters long, not the 30.48 meters Ted expected. At first, he is surprised because he thought he had used a full 100 feet (30.48 meters) of cable when he installed his antenna, but these results indicate that his buried cable is more than 3.6 meters (11.9 feet) short of that! Only then did he remember that he had used about 12 feet for an interconnection jumper required in his shack before he had buried the remainder of the cable to feed his antenna a year earlier.

Example Length2

Bob is building a new house. As he and his wife have planned for many years, he intends to have his radio equipment in the basement. He installed a cluster of 8 feet ground rods near where his equipment is to be installed before the concrete basement floor was poured.

Bob has decided on two different antennas that he plans to install in the rear part of his three acres, but he wants an additional two spare cables for possible future antennas. These four cables (Belden 7810A) are to run up in between the studs of the basement wall and exit through the side of the finished house about 3 feet above the outside ground level. Because of the shape of his basement walls, the cables need to be routed laterally about three feet before exiting the outside wall. The cables pass through specially constructed boxes that will be flush mounted at both ends. Each cable end will be terminated in a chassis connector and the exterior connector will be weather proofed.

On the weekend before the basement wall-panel installers were expected, Bob routed the four cables in between the basement studs up through the wall supports at floor level and into the outside wall of the first floor, passing through the studs and joists as required.

After the house was finished and Bob's family had moved in, Bob mounts the cable racks and connectors. It seemed like forever since he had installed the cables, but finally he is able to makeup the connections inside and outside of his house. Because he plans to measure antennas and matching networks, which were to be located outside, but intends to perform these measurements from inside of his shack, he knew that he must know the lengths of these cables, now covered and long hidden from view.

He begins by estimating the height of the basement wall and the height above ground where the outside rack is located, but he doesn't remember how far the cables were routed to the right before they continued up into the first floor wall. He finally settles on 11 feet (3.352 meters) for his estimated length.

The other known cable characteristics are:

$Z_o = 50+0i$
$v_f = 0.86$
$\alpha = 0.48$ db/100ft = 0.001813 nepers/meter. See {A-12}.

Because the estimated length is so short, it seems appropriate to use equation {9-3} to find the rough cut measurement frequency. Using equation {9-3}:

$$f = \frac{(0.135)(0.86)(2.99792 \ x \ 10^8)}{3.352} \quad = \quad 10.383 MHz$$

At this frequency, the short circuit impedance Z_{sc} is measured and found to be 2.42+110.6i ohms. However, table 9-1 shows this value for $Im[Z_{sc}]$ to be out of range. Recall that this range was established to prevent users from inadvertently making measurements that might produce errors because they were too close to the forbidden wavelength region described earlier. Bob adjusts the frequency down to 9 MHz. and repeats the measurement.

This time Z_{sc} is 1.379 +76.76i ohms, well within the guidelines. Therefore the measurement frequency was recorded at 9.0 MHz; and by using this frequency, β was calculated to be 0.21933 radians/meter.

Next, he opened Length.xls, clicks the button, enters the data, and clicks **Calculate**. Finally, Bob enters the new file name of Length2.xls and clicks **Save & Close**.

Length Transmission Line Application	
Data for:	**Length2.xls**
Input Data	
Z_o -- T. Line Char. Imped. Ohms	50+0i
Z_{sc} -- Send End Imped. Ohms	1.379+76.76i
α -- T. Line atten. nepers/m	0.0018130
β -- T. Line ph. shift radians/m	0.2193300
Output Results	
Length meters	4.5300
Length in wavelengths	0.158129

The results are shown in Length2.xls. Note that the length of the first of the four cables to be measured was found to be 4.53 meters (14.9 feet) long instead of the estimated value of 11 feet.

As an aside, note that 4.53 meters is more than 26% larger than the original length estimate of 3.352 meters. This is larger than the +120% limit for which the algorithm is designed to function when f = 10.383 MHz. However, by following the procedure which is part of Length.xls, Bob found a better measurement frequency which moved the impedance measurement frequency and assured that the application operated within its controlled region and produced the correct cable length value.

The estimate for length to be used for each of the three remaining cables should be 4.53 meters, though it is unlikely that each of them will be determined to have exactly this length.

⌘   ⌘   ⌘

10 TRANSMISSION LINE FRAGMENTS AS COMPONENTS

To this point, transmission lines have been treated as conduits, though not simple conduits, for transporting energy between sources and loads effectively. In this chapter, a descriptive analysis will show that fragments of transmission lines can also be made into impedance components, i.e., components possessing controlled real and imaginary values.

For the moment, consider fig. 6-11 from Example 6-4a.xls. The figure shows that both the real and imaginary parts of the impedance vary depending on how far from the source end of a transmission line they are measured. There is some periodicity to these variations; i.e., they behave similarly at various points along the line.

When studying fig. 6-11, note that for $x = 4.8$ meters, $Im(Z_x)$ is approximately -j150 ohms. As x increases $Im(Z_x)$ becomes equal to zero; at this value of x, $Re(Z_x)$ is at its maximum of about 300 ohms. Should x continue to increase, $Im(Z_x)$ changes sign and increases to about +j130 ohms before falling again with increasing x.

Imagine for the moment that a resistor, a capacitor and an inductor are connected in parallel. At some frequency, this simple parallel circuit would behave in exactly the same way as the transmission line in fig. 6-11 behaves at $x \sim= 4.8$ meters. Note also that in this transmission line, this behavior occurs at each of the values of $x \sim= 4.8$, 14, and 23 meters. These points are called *parallel* resonance or *antiresonance* points.

By examining fig. 6-11 at other points, namely at $x \sim= 0, 9, 18$ and 27 meters, note that $Im(Z_x)$, having been in the plus reactance or inductive range, passes through zero and becomes capacitive. In this case $Re(Z_x)$ is not at a maximum, but is instead, near zero ohms when $Im(Z_x)$ reaches

0 ohms. This is the behavior of a resistor, capacitor and inductor connected in series at resonance. Thus these points are called *series* resonance points.

These phenomena are general for transmission lines with standing waves. On the other hand, the situation described in Example 6-4c.xls as shown in fig. 6-16 has its load and source properly matched to the transmission line and thus will not provide the repeating resonant and antiresonant points along its length, because that line has no standing waves. Thus matched line fragments cannot be candidates to become reactive circuit elements. Moreover, because partially matched line fragments produce low Q resonances, they are also of minor use for our purposes here.

It follows that useful reactances using transmission line fragments should, in theory, have one end terminated in either an open circuit or a short circuit, because these terminations are the extreme opposite situation of a matched line. As a practical matter, however, open circuits are difficult to maintain in exterior environments. Even an open cable end covered by tape is not the same thing as an untaped open cable end, because the tape material will not be of the same dielectric material as the cable. Unprotected cable ends become degraded, corroded, or partially shorted over time. More importantly, with open ended cables at high frequencies there are radiation effects and undesirable coupling with other circuit elements. Therefore almost all useful fragments are terminated in a short circuit. A shorted end can be taped as tape does not change a shorted termination; and radiation and coupling effects can be minimized with short-circuited cable ends.

To develop more detail concerning the characteristics of a transmission line fragment, a reactance trial will be conducted. The first requirement is to select a cable length which is long enough for one each of both types of resonances to be observed but short enough to produce only the best resonance points, i.e., those with low values of x. Cable resonance points occurring farther from the source will be less effective because of cable losses.

As has been shown, standing waves mean that behaviors repeat every half wavelength along transmission lines. Because the cable is terminated in a short, there will be a low real impedance or series resonant point occurring at every half wavelength from the shorted load end while moving toward the source. Because parallel resonant points always occur at the mid point between series resonant points, there will be parallel resonance points at every half wavelength starting at a quarter wavelength from the load end while moving toward the source.

Reactance Trial Example

Assume that 7810-type cable, the same as that used in example 6-4, is used in this trial. Assume that the load end is short circuited, the frequency is 14 MHz, v_j is 0.86, α is 0.001813 nepers/meter, and β is 0.34118 radians/meter. Determine: 1) a reasonable cable length for the trial; 2) the approximate locations of an anti-resonant point and a resonance point on this cable, and 3) produce detailed Z_x plots around these locations.

Solution: First, from equation {3-4} determine the wavelength of the signal carried by the cable to be:

$$\lambda_{TL} = \frac{2\pi}{\beta} = \frac{2\pi}{0.34118} = 18.42 \; meters$$

As phenomena along the line repeat every half wavelength, there would be two parallel resonant and two series resonant points if a full wavelength of 18.42 meters is chosen for the cable length. This will also place a series resonance point directly at the source end of the cable. If only one parallel resonant point and one series resonant point are desired, and if no resonance point is to be located near the measurement point, select a cable length of 12 meters. Where did this answer come from?

Begin by recognizing that only one series and one parallel resonance point will occur in a cable exactly one half wavelength long; but this length will still place a series resonant point directly at the open end. Thus about a 1/8 wavelength more length is required to force that resonant point to locate away from the measurement end. That means the length should be $18.42/2 + 18.42/8 = 11.51$ meters. Rounding this, the length becomes 12 meters.

For part 2) of the problem, the parallel resonant point will occur at a quarter wavelength toward the sending end from a shorted end, making it $12 - 18.42/4 \sim= 7.4$ meters from the sending end. Because the series resonant point occurs a quarter wavelength toward the sending end from the parallel resonant point of 7.4 meters, this point will be located at $12 - 18.42/4 - 18.42/4 = 2.8$ meters.

Moving on to part 3) of the problem, open the General.xls file and enter the cable data listed earlier in the problem specifications with a length of 12 meters. To get a close look at the series resonance point at about 2.8 meters, set Startx equal to say, 2.6 meters and Stopx equal to say, 2.9 meters. Click **Calculate**, and save the file as ReactTrialA.xls. The graph produced from this file is fig. 10-1. Because 7810 is a good quality low loss cable, the value of Re(Z) will be low at the series resonance point. This resistance was calculated to be about 0.8 ohms when the reactive term passes through zero ohms. Check this in fig. 10-1.

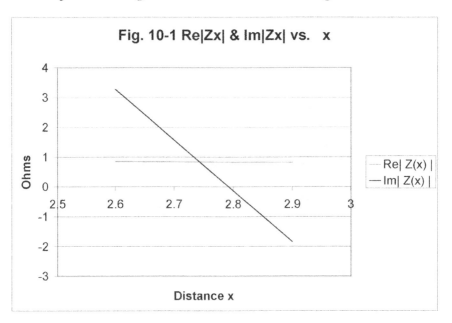

Next modify ReactTrialA.xls by changing Startx and Stopx to 7.1 and 7.7 meters, respectively, to cover the region around 7.4 meters. Click **Calculate** and produce fig. 10-2. To save the modified file, enter ReactTrialB.xls for the new name and click **Save & Close**.

Because the cable used here was the same for both resonant points, it is reasonable to expect high values of Re(Z_x) at the parallel resonance point; the graph shows Re(Z_x) to be around 5500 ohms.

Similarly, there should be high values of capacitive and inductive reactances, near the parallel resonance point; those calculated are about 3000 ohms each. From these data, it is clear that a shorted length of 7810 cable provides good resonance points and thus its fragments will produce good reactances.

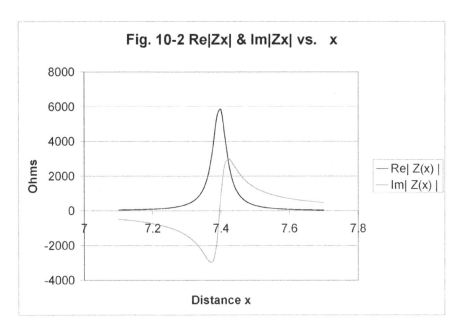

Fig. 10-2 Re|Zx| & Im|Zx| vs. x

Now that it has been established that cable fragments can produce capacitive and inductive reactances, this capability will be applied in the next two chapters.

⌘   ⌘   ⌘

11 THE SHUNTREAL APPLICATION

Given an impedance Z_f that possesses a non zero reactance at a specific frequency, assume that a fragment of transmission line, shorted at its remote end, has its open end connected in parallel with Z_f. Under these conditions, the ShuntReal application computes the length of that shorted fragment of transmission line such that the resultant parallel combination of Z_f and the shorted stub becomes purely resistive. This means that the calculation engine of ShuntReal will adjust the length of the cable fragment such that it produces a reactance value which cancels the existing reactance of Z_f.

We learned in chapter 9 that solutions for cable lengths are transcendental and require a trial and error approach. That is the situation here, because the desired unknown length cannot be expressed or written from any equation. Additionally, when one intends to proceed with this approach manually, one must start with an estimate for length, calculate a better length, reenter data, recalculate, then reenter, and recalculate until a result is reached.

However, the ShuntReal application performs these steps programmatically and requires the user to merely enter the specific problem parameters. The program automatically selects the starting value for d and improves its precision in unseen steps until a value of sufficient precision is found, all without operator intervention. ShuntReal is also stored on the recommended CD.

Example 11-1
One of your students hands you a black box that contains hidden unknown and interconnected passive components only. Its measured input impedance at 3.855 MHz is 50-j20 ohms. You have a piece of Belden 9913 cable. You desire to shunt this black box impedance with a

shorted stub made from this piece of 9913 cable such that the resulting impedance of the parallel combination, at this frequency, has only a real value. How long must the cable stub be? What is the value of the resulting real impedance of the parallel combination?

Solution: Using the methods shown in example 6-4, determine α to be 0.0007554 nepers/meter and β to be 0.096184 radians/meter. Open ShuntReal.xls, click **Enter/Change Data**, enter the required values into the input form and click **Calculate**. Enter Example 11-1.xls as the new file name, and click **Save & Close**. Example 11-1.xls then contains the required length of the shorted stub cable, its sending end impedance Z_s and the equivalent impedance Zt of the parallel combination of Z_f and the stub impedance.

The ShuntReal Application	
Data for:	**Example 11-1.xls**
Input Data	
Z_o -- T. Line Char. Imped. Ohms	50+0i
Zf -- Imped. Being shunted ohms	50-20i
α -- T. Line atten. nepers/m	0.0007554
β -- T. Line ph. shift radians/m	0.096184
Output Results	
Cable Length - meters	12.87848522
Z_s - Input Imped. Of cable ohms	4.57256671030682+144.85566099516i
Zt - Imped. Of Combination ohms	57.2767919464794-4.77906901128588E-008i
Yt - Admitt. Of Combination in S	1.74590783808985E-002+1.45675303417359E-011i

Example 11-1 Comments

As can be seen, the reactive term of the resulting parallel impedance Zt is negligible. Note that the computed length is far and away more precise than is practical; the algorithm computes the result as if all input quantities were known to this precision. Remember, however, that the third digit to the right of the decimal point of cable length has the dimensions of millimeters! It is doubtful that a physical cable of approximately 12 meters could normally be measured for use to a precision of single digit millimeters let alone to lengths even more precise.

At the frequency specified in this problem, the required cable length is nearly 13 meters. This is not a practical way to tune out 20 ohms of capacitive reactance, but at higher frequencies, stub lengths will usually be shorter, and thus the method more practical.

⌘   ⌘   ⌘

12 THE STUBMATCH APPLICATION

Unless both the load impedance and the source impedance are matched to the impedance of their inner connecting transmission line, there will be standing waves plus possible overloads and or excessive cable losses. As has been proved many times, transmission lines are most effective as energy transporters when the source and load impedances are properly matched to the line.

Matching at the source end of the line is generally not the problem because most source outputs are designed to match their impedance to the characteristic impedance of recommended transmission lines or can easily be adjusted to match them. Matching at the antenna or load, however, is a different problem. Antenna impedances are functions of type, location, size, the environment where they are used and vary widely with frequency. Example 7-2 introduced the L-C matching network as one way to match remote ends of transmission lines to their loads and there are other methods. The method used here is one of them.

There are other reasons to match load impedances to their transmission lines than just reducing VSWRs and maximizing power transfer. Unmatched transmission lines handling high power can cause insulation breakdowns and arcs that produce radio frequency interference and damage cable insulation, and standing wave voltage levels are also potentially hazardous. Thus it is desirable to make the line "flat", i.e. to minimize the standing waves on the transmission line for numerous reasons.

This chapter describes a method to make the feeder or principle transmission line flat by using a transmission line fragment. The StubMatch application is designed to accomplish this, and fig. 12-1 illustrates how the fragment, line #2, is connected to the principle transmission line, line #1.

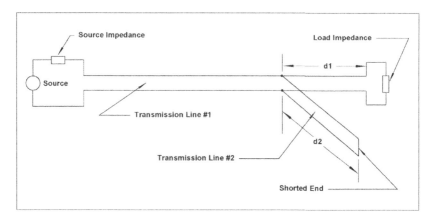

Figure 12-1 Impedance Match Using a Shorted Stub

Creating a stub match is essentially determining the lengths d1 and d2 shown in fig. 12-1.

Traditionally, the calculations necessary to accomplish a stub match are performed while assuming the transmission line to be lossless and then by employing a Smith Chart. The loss assumption is made for two reasons. When loss is taken into account, the algorithm required to perform the calculations is much more complicated than when the line is assumed to have no loss, and because the lengths of cable involved in a stub-match are usually relatively short, the errors incurred by assuming the line to be lossless are often small. For these same reasons the method used here will also assume the StubMatch cables to be lossless for the purpose of finding the lengths d1 and d2.

In general, StubMatch has other issues. While the transmission line from the source to the junction point (where three cable lengths are joined) will be made free of standing waves by using a stubmatch, neither of the other cable lengths meeting at the junction point will normally be flat. Significant standing waves may exist along their lengths; and because all transmission line reactances are frequency sensitive, the desired standing wave reduction or impedance match will work over only a limited frequency range.

The typical amateur radio operator knows well that when the antenna load impedance and the characteristic impedance of its feeder cable differ by an ohm or two then a matched situation has essentially been achieved. The StubMatch application will calculate d1 and d2 lengths, which will bring the theoretical impedance difference to within milliohms, but this assumes that the circuit and cable parameters can be determined to that precision, which is not likely. Operators must be practical and ask themselves when it makes sense to purchase several meters of cable in order to remove a few ohms of mismatch or to seek matches by using some other method.

Description of the StubMatch Application Solution Process

While the reader can use the StubMatch application without knowing how it works, the following descriptions are included for those who wish to understand more about its methods.

We assume that a transmission line is connected and matched at its source end, while its remote end is connected to an unmatched load. This means the line will have standing waves. To eliminate these standing waves from the main feeder line of fig. 12-1, a shorted stub is added. This means that the impedance "seen" by transmission line #1 at some point along its length must be made equal to the line's characteristic impedance Z_o. As will be described later, it is desirable for this junction point to be as close to the load as possible. How is the impedance level at the junction point made to become equal to Z_o?

To answer this question, recognize that the impedance varies along the length of any transmission line and line fragment where standing waves exist. Connecting two lines together at some point necessarily means that the resulting impedance at that point will be altered from its value without any connection, and by controlling *where* along each line the connection is made, the resultant impedance value at that connection point can be controlled.

As mentioned, achieving an impedance match by using a shorted stub is achieved by determining the cable lengths d1 and d2 as shown in fig. 12-1 and then connecting the stub into place. The StubMatch application first calculates the length d1 and then the length d2, in that order. As these quantities are lengths, the trial-and-error approach must be used, once more implying repetitive calculations involving complex quantities.

Like most of the applications described in this book, the Transmission Line Equation {4-1} is the starting point. Next then, we must recognize that computing quantities related to the joining of cables at the connection point means dealing with lengthy algebraic expressions of equivalent impedances and requires messy calculation of sums, products and quotients of these complex expressions, etc. Because the use of admittances instead of impedances simplifies these types of calculations considerably, StubMatch employs admittances internally.

Converting impedances to admittances for these expressions, though lengthy and difficult when done algebraically, is conceptually just $Y = 1/Z$. Additionally, as the algorithm needs to work for virtually any cable characteristic impedance Z_o, all admittances Y are normalized. Impedances are normalized by dividing the impedances by Z_o while admittances are normalized by multiplying them by Z_o.

Making the principle transmission line, i.e., transmission line #1 in fig. 12-1, see its own characteristic impedance Z_o at the junction point is equivalent to having that line "see" its own characteristic admittance Y_o at that point. Because $Y = g + jb$ is complex, achieving the desired solution means that the *real* part of the junction point normalized admittance, i.e., its conductance g, must approach a numerical value of 1 (one) as the length d1 is varied.

Thus we use a trial-and-error approach requiring repetitive calculations of the Transmission Line Equation, in the modified form we have just discussed, while varying d1. This continues until the value of the conductance g is forced to approach a value of 1. Once the conductance g is sufficiently close to a value of 1, that value of length which enabled g to become equal to 1 will be the value of d1 we seek. Having found d1, the normalized susceptance b associated with the g of this d1 solution becomes the key to finding d2.

StubMatch then computes d2, the length of a shorted stub fragment, which, by adjusting its length, produces a normalized susceptance equal to the negative of the b found in the previous paragraph by varying the length d2. With these two equal but oppositely signed values of b the resultant total normalized susceptance at the junction will equal zero, leaving only a real value of normalized admittance $Y = g$, which equals 1. Thus we have determined d1 and d2 by forcing g to become equal to a value of 1 and by eliminating the imaginary part of Y.

The next step is to convert this admittance to an impedance, i.e., $Z = 1/Y = 1/g$ yielding $Z = (1/1)$; finally, we remove the normalization by multiplying this Z by Z_o, i.e., $(1/1)(Z_o) = Z_o$. Thus the impedance value at the junction point now equals the cable's characteristic impedance. With these actions then, standing waves have been eliminated from the main transmission line #1 for the frequency of interest.

As has been demonstrated several times in this book concerning transmission lines, electrical behaviors at a fixed frequency, repeat every half wavelength along their lengths; thus our transmission line will be flat when the length is d1, and length d1 + $\lambda/2$, and d1 + λ, and d1 + $3\lambda/2$ etc. Because every d1 requires a d2, the d2 values will also repeat for each half wavelength. In general then, there are, potentially, many points on line #1 where a fragment could be connected and many stub lengths which can make line #1 flat. StubMatch was designed to find that junction point which is closest to the load, resulting in the shortest possible d1 and shortest possible stub section d2. This minimizes losses, maximizes that portion of the main cable which is flat and minimizes the total cable lengths possessing standing waves.

Users attempting to use a calculator with this description to find d1 and d2 should sketch out each step on a Smith Chart. In this way, this rather complicated-sounding process can be kept more clearly in mind. This suggestion is especially important for those cases where the length d1 passes through the mathematical discontinuity posed by the quarter wavelength point on the transmission line. Should this event be overlooked, the solution for the length d1 can easily be in error. This is true because the desired length may consist of two pieces of length added together. This will be clear when looking at a Smith Chart but it is not so obvious without one.

Example 12-1

In fig. 12-1, say that the characteristic impedances of each line is Z_o = 100+j0 ohms and the load impedance Z_L is 150+j50 ohms. Find the values for d1 and d2 such that the main transmission line "sees" its own characteristic impedance at the junction point. Example taken from Kraus. (Kraus 1953: 436)

Solution: Kraus uses a normalized value of β = 1.0 for this problem. This removes the need to specify the operating frequency; but equation {3-4} can always be used to determine the wavelength. Note that Kraus assumes that α is equal to zero as discussed in the text. Enter β, Z_o and Z_L into a StubMatch application's input form; then click **Calculate**. Enter the new file name of Example 12-1.xls and click **Save & Close**. Example 12-1.xls shows the results.

The StubMatch Application	
Data for:	**Example 12-1.xls**
Input Data	
Z_o -- T. Line Char. Imped. Ohms	100+0i
Z_L -- Rec. end load Imped. Ohms	150+50i
β -- T. Line ph. shift radians/m	1
Output Results	
d1 distance to junctn from load	1.219917
d1 in wavelengths	0.194156
d2 length of stub	1.047198
d2 in wavelengths	0.166667

Kraus's results were obtained using a Smith Chart and were:

d1 = 0.194 wavelengths

d2 = 0.167 wavelengths

Example 12-1 Comments:

It can be proved that adding the stub described above does indeed eliminate standing waves on the line from the source to the junction point. To check this, assume that each of the three transmission line sections is disconnected from the junction point without changing any lengths. See fig. 12-2.

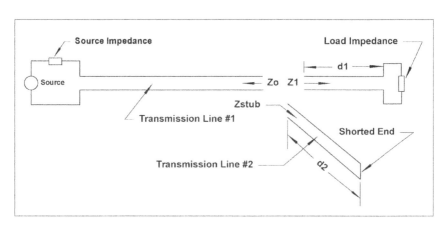

Figure 12-2 Stub Match with Elements Separated

Next, apply the General application to the d1 section of cable to find its sending end impedance. This requires that $Z_L = 150+50i$, $Z_o = 100+0i$, $\alpha = 0$, $\beta = 1.0$ and length = 1.219917 meters, which was obtained in Example 12-1.xls, all be entered into the General application data input form. Click **Calculate**, enter ExampleZ1.xls as the new file name, and then click **Save & Close**. See Example Z1.xls. Note that Example Z1.xls is also stored on the recommended CD for reference.

Repeat these actions for the stub portion of fig. 12-2. In this case, the inputs are $Z_L = 0+0i$, $Z_o = 100+0i$, $\alpha = 0$, $\beta = 1.0$, and length = 1.047198 meters. The file is saved as Example Zstub.xls. This file is also included on the CD.

Because the load for transmission line #1 is the parallel combination of the sending impedances from Z1 and Zstub, this parallel combination should be computed next. This parallel impedance is $Z\| = (Z1*Zstub)/(Z1 + Zstub)$, and the easy way to calculate this is by using Excel's complex functions. This was done using cells D17 through E21 of the InOut1 sheet of ExampleZstub.xls. The results do not appear on the version of ExampleZstub.xls as shown below, but they do appear in table 12-1 and on the stored copy of ExampleZstub.xls on the CD.

This calculation shows the reactive term to be negligible, thus $Z\| = 100+0i$ ohms. The use of the stub forces the impedance at the junction point to become equal to Z_o and thus transmission line #1 has been made 'flat' by using the stub fragment.

The General Application	
Data for:	**ExampleZ1.xls**
Input Data	
Z_o -- T. Line Char. Imped. Ohms	100+0i
Z_L -- Rec. end load Imped.Ohms	150+50i
d -- T. Line length meters	1.219917
α -- T. Line atten. nepers/m	0
β -- T. Line ph. shift radians/m	1
V_g -- Send.end source voltage	1+0i
Z_g -- Send.end source imped.	0
Startx	0
Stopx	0
Output Results	
K_r -- Refl. Coef. at rec. end	0.230769230769231+0.153846153846154i
abs| K_r |	0.277350098
K_s -- Refl. Coed. at send end	-7.69231217311563E-002-0.266469342075614i
abs| K_s |	0.277350098
$VSWR_s$ -- VSWR at send. end	1.767591879
$VSWR_r$ -- VSWR at rec. end	1.767591879
Z_s -- send. end imped.	74.9999945390158-43.3012649343867i
Y_s -- send . end Admittance	1.00000009708418E-002+5.77350297215405E-003i
I_s -- send. end current	1.00000009708418E-002+5.77350297215405E-003i
abs| I_s |	0.011547006
I_r -- rec. end current	3.43723723748977E-003-7.40621855687227E-003i
abs| I_r |	0.008164966
E_s -- send. end voltage	0.999999999999999+2.22044604925031E-016i
abs| E_s |	1
E_r -- rec. end voltage	0.885896513467079-0.939070921656352i
abs| E_r |	1.290994511
P_g -- Total gen/source Pwr.	0.013333334
P_{sr} -- Pwr used by source	0
P_s -- Pwr supplied to T. Line	0.010000001
P_r -- Pwr consumed by load	0.010000001
P_L -- Pwr consumed by T. Line	2.94903E-17

Thus with the addition of the stub, the equivalent load connected to transmission #1 is a real resistance and is able to absorb power. This is true because the reactive currents within the load components are equal but opposite in phase, leaving only real currents and producing a real impedance which equals the line impedance.

The General Application

Data for:	ExampleZstub.xls
Input Data	
Z_o -- T. Line Char. Imped. Ohms	100+0i
Z_L -- Rec. end load Imped.Ohms	0+0i
d -- T. Line length meters	1.047198
α -- T. Line atten. nepers/m	0
β -- T. Line ph. shift radians/m	1
V_g -- Send.end source voltage	1+0i
Z_g -- Send.end source imped.	0
Startx	0
Stopx	0
Output Results	
K_r -- Refl. Coef. at rec. end	-1
abs$\mid K_r \mid$	1
K_s -- Refl. Coed. at send end	0.500000777350094+0.866024954980687i
abs$\mid K_s \mid$	1
$VSWR_s$ -- VSWR at send. end	6.0048E+15
$VSWR_r$ -- VSWR at rec. end	#DIV/0!
Z_s -- send. end imped.	173.205260278389i
Y_s -- send . end Admittance	-5.77349670785242E-003i
I_s -- send. end current	-5.77349670785242E-003i
abs$\mid I_s \mid$	0.005773497
I_r -- rec. end current	-1.15470023917717E-002i
abs$\mid I_r \mid$	0.011547002
E_s -- send. end voltage	1
abs$\mid E_s \mid$	1
E_r -- rec. end voltage	0
abs$\mid E_r \mid$	0
P_g -- Total gen/source Pwr.	#DIV/0!
P_{sr} -- Pwr used by source	0
P_s -- Pwr supplied to T. Line	0
P_r -- Pwr consumed by load	0
P_L -- Pwr consumed by T. Line	0

```
Z1=       74.9999945390158-43.3012649343867i
Zstub=    173.205260278389i
Znum=     7500.00686334393+12990.393575008i
Zden=     74.9999945390158+129.903995344002i
Z || =    99.9999902915439-6.26430040386349E-005i
```

Table 12-1 Parallel Impedance Calculations

Example 12-2

Albert and a number of his buddies have talked to each other on 80 meters for years. Almost every evening, they congregate around 3.65 MHz and chew the rag. Sometimes the band is busy and they move up or down in frequency but they are usually somewhere between 3.6 MHz and 3.7 MHz. Recently, Albert changed his residence and his new homeowners association rejected his plan to put up his old 80 meter dipole antenna.

After doing some computer modeling, Albert proposed a single vertical rod about 30 feet tall, insulated from ground and held in place with guys attached to the rod about 15 feet up and at its top end. The opposite ends of the guys would be fastened to ground anchors. The committee liked his proposal and was ready to approve it. When he told them he might have five guys at each level and that the guys might be made of Dacron or be made of a conducting material, the committee members said they didn't care. They believed that his antenna would hardly be noticed and approved it.

Construction began. He made the vertical rod from 1-inch copper pipe with a ¾ inch iron pipe welded inside for strength and stiffness. He decided to use 5 guys at each level; this meant that the ground anchors would be 72 degrees apart around the base of the antenna. The lower level guys would be made of Dacron but the top guys would be copper-weld wire mechanically and electrically attached at the top but extending only about half way down toward their anchors. The other half of each of the top guys would be Dacron connected to the copper-weld wires via insulators and running the remainder of the distance to the ground anchors.

The model of such an antenna indicates that it would produce a low angle, omnidirectional radiation pattern and would perform better with a good set of ground radials. The modeling predicted one problem: the antenna's driving point impedance would be low enough to cause a high VSWR on a 50 ohm transmission line.

- Question 1: For such an antenna, what is the actual VSWR at 3.6, 3.65 and 3.7 MHz?
- Question 2: What are the d1 and d2 lengths which will produce a stub match at 3.65 MHz?
- Question 3: What are the VSWRs for all three frequencies when a stub match calculated at 3.65 MHz is attached to his feeder cable near his antenna?

Solution for Question 1: To determine these VSWRs, the antenna input impedance at these three frequencies must be known. Albert's most accurate signal source and impedance bridge require AC power and both instruments were located near his transmitter inside. Because the equipment's presence at the base of the antenna would affect any measurements he might make there, Albert decided to make his impedance measurements at the transmitter end of a transmission line next to his source and bridge. After connecting the remote end of a 25 meter length of Belden 7810A cable to the base of his antenna, he set his source frequency, in turn, to 3.6, 3.65 and 3.7 MHz and recorded the three impedance measurements. See the Z_s column in table 12-2.

His next move was to use the Reverse application, but first he had to determine his cable parameters. The cable length was given as 25 meters, β was determined by using equation {3-3} for each frequency. By using the cable data tables for Belden 7810A cable and equation {A-12}, α for the 25 meter cable was calculated to be 0.000755 nepers/meter. The Reverse application was used three times, once for each frequency, to determine the three antenna input impedances. See the Z_{ant} column in table 12-2.

Freq. MHz	Meas. Cable Input Zs	β	Zant	VSWR
3.6	28.69+117.5i	0.087733	11.37-77.94i	11.85
3.65	52.53+146.4i	0.088952	11.76-68.56i	10.06
3.7	113.2+182i	0.09017	12.17-59.23i	8.44

Table 12-2
Measurements and Calculated Values

As a result of these actions, Albert had the cable length, the attenuation, the phase shift, and the load impedances necessary to use the General application to find the VSWRs without any matching network in place. The General application requires a non-zero source voltage. For this, he assumed $1+0i$ volts, though any nonzero voltage could have been used. After running the General application three times, once for each frequency, he was able to record the three sending end VSWRs. These results are shown in the VSWR column of table 12-2.

The StubMatch Application	
Data for:	**Example 12-2.xls**
Input Data	
Z_0 -- T. Line Char. Imped. Ohms	50+0i
Z_L -- Rec. end load Imped. Ohms	11.76-68.56i
β -- T. Line ph. shift radians/m	0.08895
Output Results	
d1 distance to junctn from load	7.566937
d1 in wavelengths	0.107124
d2 length of stub	3.368079
d2 in wavelengths	0.047681

Solution to Question 2: Belden's 7810 cable has a $v_f = 0.86$, table 12-2 provides the β for the StubMatch frequency of 3.65 MHz, and the antenna driving point impedance for 3.65 MHz comes from table 12-2. After entering these data into the StubMatch application data input form, he clicks **Calculate**, enters Example 12-2.xls as the new file name, and clicks **Save & Close**. The lengths for d1 and d2 are listed there for a frequency of 3.65 Mhz.

Solution for Question 3: There are then two ways to determine the desired VSWRs with the stub match cable connected in place: The first method is to simply connect the matching stub and then measure and record the VSWR at each of the three frequencies. The second way is to calculate the VSWRs. Albert decided to calculate them.

This meant that Albert must know the equivalent load impedances for each of the three frequencies. Fig. 12-2 shows the pieces of a StubMatch broken apart. From these pieces, the impedances Z1 and Zstub can be computed for each frequency by using the General application, as was done for example 12-1. Remember, however, that all of the Z1 calculations must use 7.566937 meters for the cable length; likewise all of the Zstub calculations must use 3.368079 meters as they were determined from Example 12-2.

The results of these calculations for the three values of Z1 and Zstub are shown in table 12-3. Again, as in example 12-1, these two impedances in parallel form the load for the transmission line. Each of the three parallel impedances were computed in complex form for their respective frequencies and appear in the $Z\|$ column of table 12-3. The final step is to use the General application three times, once for each value of the equivalent load impedance of $Z\|$ shown in table 12-3, length as 25 meters, α as 0.000755 nepers/meters and the appropriate β from table 12-3 to compute the three VSWRs. The results appear in table 12-3.

Freq. MHz	β	Cal. Using Reverse Ap. Zant	Cal. Using General Ap. Z1	Zstub	$Z\|$ = Z1*Zstub/(Z1 + Zstub)	VSWR
3.6	0.087733	11.37-77.94i	3.697-17.79i	15.22i	42.26+44.58i	2.48
3.65	0.088952	11.76-68.56i	4.355-14.10i	15.44i	50.0+0.01183i	1
3.7	0.09017	12.17-59.23i	5.193-10.00i	15.67i	21.59-7.881i	2.3

Table 12-3 Calculations leading to VSWRs

Only the final results are shown in the tables for this example, but readers are encouraged to work through the intermediate steps for themselves.

Example 12-2 Comments

Note that the calculations used to find the StubMatch lengths assumed that d1 and d2 were lossless, but one should not assume lossless transmission lines when performing calculations involving the principle transmission line. Therefore, in the above calculations, the principle transmission line was not assumed to have zero attenuation for the VSWR and impedance calculations. Instead the specified α given for the 7810 cable was used.

Table 12-4 shows the VSWRs versus frequency without a stubmatch compared to the VSWRs with a stubmatch. The table shows that there is an obvious improvement by using the stubmatch. Note that with the stubmatch in place, Albert's antenna has VSWRs close to 2.0 for the entire sub band of interest. This means that one would expect near maximum radiated power would be achieved if a quality transmission line and a transmitter tuner are used.

The fact that Albert's bare-bones antenna, without any attached matching, had such high VSWRs is an indication that the antenna's performance might be and perhaps should be improved before applying any matching scheme. If this were to be done, it is possible that his antenna might perform as he wants without any matching or at least perform better should matching be eventually applied. Possible adjustments to improve his bare-bones antenna might include the use of three or four or even six guys instead of the five specified. Perhaps the conductive portions of the top guys could be lengthened or shortened, or other adjustments which might raise the driving point impedance or make it less reactive. Any of these actions might lead to an antenna which requires less matching than the one described and possibly even increase the amount of power radiated by the antenna and or improve its bandwidth.

Operating frequency	VSWR w/o stub match	VSWR with stub match
3.6 MHz	11.85	2.48
3.65 MHz	10.06	1
3.7 MHz	8.44	2.3

Table 12-4   VSWR
Comparisons

It should be pointed out that the problem specification was concerned only with the standing wave ratios for the proposed antenna. However, almost any other performance measure could have been selected and calculated by using these applications to determine the antenna's suitability for use. For example, antenna power, antenna current, antenna voltage, transmission line loss, reflection coefficients, cable driving point impedance and others could have been used to characterize the antenna, and each one could have been calculated by one or more of the applications described in this book.

Note that in addition to the StubMatch application, which is the topic of this chapter, this example used a number of the other applications and in fact required their repeated use. As a result, it is hoped that the reader is able to see that various applications are and can be used to obtain solutions for many more things than those that are most obvious. In addition, when applications are used repeatedly, the one or two new data entries for each click of **Calculate** are easily and quickly made without the need to reenter the fixed common data, thus speeding the data entry and calculation process.

⌘ ⌘ ⌘

13 SUMMARY

An electrical circuit consisting of a power source connected to a load impedance by using a transmission line is said to be characterized and understood when numerical values for input and output voltages, currents, powers, reflection coefficients, standing wave levels, etc. are known or can be calculated. The solution to the Transmission Line Equation enables all of these quantities and more to be calculated.

The book discusses the line parameters involved and leads the reader through an outline of the mathematics related to this equation. The reader is free to seek numerical results to transmission line problems in two ways: apply the mathematical equations directly, using function tables and a computational device or use one or more of the six preprogrammed computer applications that were developed and constructed specifically for this book. When the second choice is selected, detailed knowledge of the applicable mathematics is not required in order to obtain useful numerical results for practical problems.

Numerous mathematical relationships which are part of the understanding of transmission lines are presented first, where it becomes clear that the transmission line's inherent parameters directly affect calculated results. This is followed by a short chapter which introduces six easy-to-use computer application programs and provides simple instructions concerning their use. The book then moves to the solutions of examples using the application programs.

Because knowledge of the transmission line's inherent parameters is required for problem solution, two chapters address how to determine these parameters. One of the methods uses physical dimension measurement and estimates, another employs electrical measurements, and another discusses how to use the data supplied by transmission line manufacturers. A method describing how to determine the transmission line length when the line is not accessible for physical length measurement is also presented.

The General application computes many electrical quantities pertaining to the performance of transmission lines and their loads. Numerous examples are presented to illustrate how varying certain transmission line parameters affect voltages, currents, impedances, VSWRs, reflection coefficients, various powers, etc. Using still more examples, one learns the effects of impedance mismatches, line losses, and just what transmitter tuners can and cannot do to improve performance of transmission line systems. Thorough explanation follows the solution of each example and observed behavior.

The Reverse application assumes that the sending impedance is already known but that the load impedance is not known, and thus it becomes the quantity being sought. This chapter and application shows how easily unknown load impedances can be determined. This chapter also includes the first example of impedance matching at the load end of a transmission line. The method employs an L-C network. A derivation of the equations required for calculating the element values for such a network is also shown in Appendix B. The effectiveness of a matching network using the L-C network is thoroughly analyzed and discussed.

Next, the characteristics of transmission lines and transmission line fragments are discussed; methods are given which predict where along a transmission line's length certain behaviors occur. The fifth application presents an example where the reactance of a line fragment is used to eliminate the reactance of a connected impedance and to turn the parallel combination of the load and the line fragment into a real valued impedance.

The last chapter demonstrates how to use a line fragment to provide an impedance match of a load to its feeder transmission line. The mathematics and steps used to accomplish this are fully described, and examples are presented and discussed.

Because various units of measure are used in transmission line work, Appendix A supplies formulas for conversion of a few quantities and also presents methods for derivation of units conversion factors.

⌘   ⌘   ⌘

APPENDIX A CONVERSION OF UNITS OF MEASURE

While versions of the so-called metric system of units existed in the early part of the eighteenth century, it was not until post-revolutionary France adopted such a system that it received wide spread recognition. Though Thomas Jefferson argued for adoption of the metric system as well as a system of decimal money for newly formed America, only the decimal money system was adopted, and the English system of measurement units came to America. The irony is that the English have since abandoned both their non-decimal currency system and their non-decimal system of measurement units, leaving Myanmar, Liberia and the United States as the only countries in the entire world still officially using the English system of units. However, the fields of science, medicine and engineering, moved to the metric system long ago, and virtually all scientific and medical exploration, research and analysis use the metric system. But people cling to familiar things and thus the clumsy English system remains.

No small part of the ungainly calculations required when using the English system is due to its use of so-called vulgar or common fractions (see example A-1). However, some method for keeping track of the leftover or remainder portions of calculated results was and is required. Complications wrought by the old English system were also created in part by their use of irregularly related dimensions (for example, 12 inches per foot, 3 feet per yard, 5280 feet per mile etc).

There is more to the story. While history reports that base 10 or decimal fractions were first used by the Chinese during the first century BC, decimal fractions did not come into common use until well after the English began counting and measuring. By then, the English had developed a system of inverse powers of 2 to represent their left-over portions of whole quantities; i.e. they had begun to use ½, ¼, 1/8, 1/16, 1/32 etc., of a whole portion for their remainders. Consequently, these common fractions became part of and remain a part of the English System of units today.

Conversion of many fractional quantities is simple and easy but not all of them are obvious. An illustration of a not so simple conversion is demonstrated next in Example {A-1}.

Example A-1

A cable is 27.2 meters long; how long is this cable in feet and inches?

Solution: Use the factor that converts meters into decimal inches (see this factor below in Equation {A-7}).

$$length_{inches} = \left(39.37007874 \ \frac{inches}{meter} \right) \ x \ (27.2 meters)$$
$$= 1070.866142 \ inches$$

Continuing with the conversion, find the number of whole feet in this length:

$$whole \ feet = (1070.866142 \ inches) \ x \ \frac{1 \ feet}{12 \ inches}$$
$$= 89 \ feet$$

Then find the difference between the total length and the length in whole feet:

Diff. in inches = 1070.866142 - (89) * (12)
$$= 1070.866142 - 1068$$
$$= 2.866142$$

Thus we know that our length is 89 feet, plus 2 inches, plus a left-over quantity which is less than one inch.

When dealing with a number with a value less than one in the English system of units, that remainder is normally expressed as a fraction of an inch; thus our task is to decide how much precision is required. This means our desired result could be:

$$0.866142 \ inches \ \sim= \frac{1.7}{2} \ inches$$

or

$$0.866142 \ inches \ \sim= \frac{3.5}{4} \ inches$$

or

$$0.866142 \ inches \ \sim= \frac{6.93}{8} \ inches$$

or

$$0.866142 \ inches \ \sim= \frac{13.9}{16} \ inches$$

or

$$0.866142 \ inches \ \sim= \frac{27.7}{32} \ inches$$

or

$$0.866142 \ inches \ \sim= \frac{55.4}{64} \ inches$$

Obviously this series could be continued, but if we stop here our 27.2 meter cable expressed in English units is approximately

$$= 89 \text{ feet 2 and } 55/64 \text{ inches in length.}$$

This exercise should make clear the benefits of decimal calculations when compared with a system which uses fractions.

Units Conversion Introduction

Independent of whichever numerical counting system is used or how it handles its remainder, mankind has developed numerous measurement systems; the English and metric systems are two of many. It is to be expected that quantities characterized by length, mass, temperature, power, etc. will be expressed in many of these measurement systems. And as technologies have evolved these measurement systems have also developed. Thus it is natural that quantities continue to be defined in various systems of measurement. Consequently, there is a need to convert quantities expressed in one system of units into other systems of units. It is our aim to show how units conversions are performed and to develop a few often-used conversion factors.

Unit conversions involve the use of algebra. Some rules of algebra become so automatic that users sometimes forget their importance. Because one or two of these rules may be used in the units conversion process, perhaps in unfamiliar ways, we will review a few of them.

It is important to remember that these rules are required to maintain the integrity of any and all algebraic equations and terms of an equation. The rules important to us are as follows:

1. When a non zero quantity is added to or subtracted from one side of an equation, the same quantity must also be added to or subtracted from that equation's second side.
2. When one side of an equation is multiplied by a finite quantity the second side must also be multiplied by the same quantity.
3. When one side of an equation is divided by a non zero quantity, that equation's second side must also be divided by the same quantity.

12 inches = 1 foot
1 volt = 1000 millivolts
2.54 cm = 1 inch
1 meter = 100 cm
1 horsepower = 745.69987 watts
1 joule = 1.05587 Btu

Table A-1
Several Units Conversions

An important point related to these rules, is:

Any arithmetic or algebraic term may be multiplied by or divided by the quantity one (1) without affecting the integrity or use of that term.

As stated earlier, the physical world has invented differing units of measure for many quantities. This means that a large number of equations for every conceivable unit of measure of the form shown in table A-1 have been or can be written down to make units of measure conversions. Remember that these equations always relate only measures of the same type to one another. For example, they only relate one measure for length to another measure for length, one measure of power to another measure of power, one measure of mass to another measure of mass etc. There can be no legitimate equation relating a measure for length purely to a measure for power, for example.

To use the relationships defined by the equations in table A-1, two additional operations are required. The first operation is to divide both sides of each of the equations in table A-1 by the equation's right side. The second is to divide each equation in table A-1 by its left side. These actions yield the expressions in table A-2.

$$\frac{12 \text{ inches}}{1 \text{ foot}} = 1 = \frac{1 \text{ foot}}{12 \text{ inches}}$$

$$\frac{1 \text{ volt}}{1000 \text{ millivolts}} = 1 = \frac{1000 \text{ millivolts}}{1 \text{ volt}}$$

$$\frac{2.54 \text{ cm}}{1 \text{ inch}} = 1 = \frac{1 \text{ inch}}{2.54 \text{ cm}}$$

$$\frac{1 \text{ meter}}{100 \text{ cm}} = 1 = \frac{100 \text{ cm}}{1 \text{ meter}}$$

$$\frac{1 \text{ horsepower}}{745.69987 \text{ watts}} = 1 = \frac{745.69987 \text{ watts}}{1 \text{ horsepower}}$$

$$\frac{1 \text{ joule}}{1.05587 \text{ Btu}} = 1 = \frac{1.05587 \text{ Btu}}{1 \text{ joule}}$$

Table A-2 Conversion Ratios

Note that we now have two columns of quantities and all of the quantities involved are equal to one (1) despite the quantities having different units. This means that any term may be multiplied by any of the quantities within the two columns and we will not have injured the legitimacy of the original quantity. By this action, the units of the original term will have been changed.

Keep in mind that the equations listed in table A-2 are only a sample of the equations which might be written and used, but all of them will be of the form shown: conversion ratios and reciprocals of conversion ratios both possessing the value of one (1).

The Fundamental Conversion Process

We shall now demonstrate the fundamental conversion process by using simple examples.

Example A-2

Say that we posses an item which is 6.8 feet in length and we desire to find that item's length expressed in decimal inches. This means that we must find a conversion factor from table A-2 or similar table, which involves both feet and inches and only feet and inches. Thus we look to find those conversion ratios that involve both of these two units. The first line in table A-2 fits our needs. Because both of the items in the first row involve feet and inches it is clear that there are two choices. If we were to multiply 6.8 feet by the conversion ratio in the second column, we get:

$$6.8 \ feet \quad x \quad \left(\frac{1 \ feet}{12 \ inches} \right)$$

$$= \ 0.566667 \ \frac{feet^2}{inches} \hspace{3cm} \{A-3\}$$

This multiplication has not compromised the integrity of the 6.8 feet term, but this is not the result we desire, because this result is not a length expressed in inches alone. Consequently, we try the conversion ratio from column one instead:

$$6.8 \ feet \ x \ \left(\frac{12 \ inches}{1 \ feet} \right)$$

$$= \ 81.6 \ inches \hspace{3cm} \{A-4\}$$

and get the result we want. Note that the "feet" associated with our original 6.8 feet quantity is canceled by the "feet" in the correct conversion ratio when the multiplication is carried out. This test should be applied when using any conversion ratio. Ask yourself, "Do the units properly cancel?" If they do not, the conversion ratio is incorrect or incomplete and most likely was selected from the wrong column for the problem at hand.

Example A-3

Assume that we have a length equal to 7.8 feet and want to express it in cm (centimeters). If we study the equations in table A-2 we quickly realize that there are no entries that have only feet and cm for units. This means that we have to search other tables and charts to find the desired conversion ratio or in this case we can use more than one conversion ratio from table A-2 to produce the units we desire; i.e., we could convert feet to inches with one factor and then use a second factor to convert inches to cm, as in:

$$7.8 \; feet \quad x \quad \left(\frac{12 \; inches}{1 \; feet} \right) \quad x \quad \left(\frac{2.54 \; cm}{1 \; inches} \right)$$

$$= \; 237.744 \; cm$$

Because each conversion ratio is equal to one (1), there can be as many conversion ratios included as required to reach the desired unit. Make sure that the units properly cancel throughout the ratio string, that the proper numbers are used, and that all of the required multiplication and division is carried out.

Most units conversions are this simple: locate the proper conversion factor, multiply the original quantity by the selected conversion factor and, check that there has been the necessary units cancelation. Other conversions may not be as straight forward as these examples.

Attenuation/Gain Conversions

The two principal units used when discussing attenuation and or gain in the electrical world are the neper and the decibel. Both are based upon logarithms of ratios. The neper was chosen as a unit to honor John Napier, the inventor of logarithms. The bel was chosen to honor Alexander Graham Bell, America's telephone pioneer, but it was found to be of an inconvenient size, and the decibel or db came into use. By using the decibel instead of the bel, the magnitudes of numbers associated with its calculations became 10 times larger.

123

The neper is based upon the natural logarithm that uses the base e = 2.71828... while the decibel uses the common logarithm with a base of decimal 10.0. Neither of these ratios are part of the metric system of measurements or of the English system. Ratios are dimensionless after all, but the neper is most often used with the metric system, which is why the Transmission Line Equation solution used the neper. Nevertheless the decibel is used widely, especially in the United States, prompting the need for conversions between the two units.

To develop a method ultimately leading to conversion factors for the decibel and the neper, it is necessary to take an extra step. Assume that the output voltage or current taken from a given network is two times larger than the input voltage or current applied to that network. (The number two is arbitrary; any number can be used.) Thus the gain produced by this assumed network expressed in db is

$$db \; = \; 20 \log(2)$$

and the gain expressed in nepers for this same network would be:

$$nepers \; = \; \ln(2)$$

Because the network is the same in both cases, the gains or losses specified by each system of units must also equal to each other; thus:

$$20 \log(2)db \; = \; \ln(2)nepers$$

Carrying out the mathematics indicated by this equation:
20(0.30102999566398) db = 0.69314718055995 nepers
6.02059991327962 db = 0.69314718055995 nepers

By dividing as indicated above in order to make both conversion factor quantities equal to one (1):

$$0.1151292546\ 497 \; \frac{nepers}{db} \; = \; 1$$

$$= \; 8.6858896380\ 6504 \; \frac{db}{nepers} \qquad \{A-5\}$$

Thus we now have the two conversion ratios for attenuation and or gain.

Example A-4

Using calibrated instrumentation, an amateur radio operator measures a change in signal voltage of 7.6 db after adjusting a network at the base of his antenna. What is the equivalent of this change expressed in nepers?

$$change \quad = \quad 7.6 \ db \ \ x \ \ 0.1151292546 \ 497 \ \frac{nepers}{db}$$

$$= \ 0.87498 \ nepers$$

Conversions When Quantities Possess Multiple Units

Until this point, the discussion has been about conversion of units which involve only one quantity type, i.e., length or gain or whatever. Conversions in general, however, will involve multiple quantities. A case in point of special interest to those studying transmission lines will illustrate the issue.

The following discussion will center on understanding and converting quantities with multiple units by way of examples. In each case, the subscript of variables represent the units for that variable. For example, L_{inches} means that the quantity L is expressed in inches.

Transmission line attenuation in the United States is frequently specified in db/100 ft. Note that this unit involves two types of quantities: gain or attenuation and length. In addition, the length quantity is perverted, in that the length quantity also includes a specific constant of 100. This situation is corrected by removing this numerical constant. Should a value for attenuation be specified in db/100 feet, this attenuation can be converted to db/feet as follows:

$$\alpha_{db/ft} \quad = \quad \frac{\alpha_{db/100ft}}{100}$$

By dividing a specified attenuation expressed in db/100 feet by 100 and then substituting the result into any equation which specifies the attenuation in db/100 feet, converts the attenuation into a specification expressed in db/feet, a result without any constant.

Example A-5

The Belden Company offers transmission line using their number 8237 of type RG-8, which has a specified attenuation of 0.6 db/100 ft at 10 MHz. Find the conversion ratios and the attenuation of the cable expressed in nepers per meter at this frequency.

Solution: It is clear that there will be at least two conversion ratios involved (one for length and one for attenuation), but in general more ratios could be required. Should we obtain our ratios from table A-2, the length conversion will require three ratios:

$$Length\ conv\ =\ \left(\frac{1\ ft}{12\ inches}\right)\left(\frac{1\ inches}{2.54\ cm}\right)\left(\frac{100\ cm}{1\ meter}\right)$$

The gain factor is the single term from {A-5}:

$$gain\ /\ loss\ =\ \left(\frac{0.1151292546\ 49702\ nepers}{db}\right)$$

The answer to the example is composed of three factors affecting length and one factor affecting attenuation all multiplied by the given value of $0.6_{db/100\ ft}$, which has also been divided by 100 as described earlier, i.e.:

$$\alpha_{nepers\,/\,meter}\ =\ \left(\frac{0.6}{100}\right)\left(\frac{db}{ft}\right)x\left(\frac{1\ ft}{12\ inches}\right)x\left(\frac{1\ inches}{2.54\ cm}\right)x\left(\frac{100\ cm}{1\ meter}\right)x$$
$$\left(\frac{0.1151292546\ 49702\ nepers}{db}\right)$$

This yields results for α in nepers/meter of:

$$\alpha_{nepers\,/\,meter}\ =\ 0.0022663239\ \frac{nepers}{meter}$$

Thus at 10 MHz, Belden 8237 has an attenuation which is: 0.6 db/100 ft or equivalently, an attenuation of 0.0022663239 nepers/meter.

Frequently Used Units Conversion Equations

Inches to Meters

$$length_{meters} = 0.0254 \frac{meters}{inches} \quad x \quad length_{inches} \qquad \{A-6\}$$

Meters to Inches

$$length_{inch} = 39.37007874 \frac{inches}{meter} \quad x \quad length_{meter} \qquad \{A-7\}$$

Feet to Meters

$$length_{meters} = 0.3048 \frac{meters}{feet} \quad x \quad length_{feet} \qquad \{A-8\}$$

Meters to Feet

$$length_{feet} = 3.280839895 \frac{feet}{meters} \quad x \quad length_{meters} \qquad \{A-9\}$$

Decibels to Nepers

$$gain_{nepers} = 0.1151292546\ 497 \frac{nepers}{db} \quad x \quad gain_{db} \qquad \{A-10\}$$

Nepers to Decibels

$$gain_{db} = 8.6858896380\ 6504 \frac{db}{neper} \quad x \quad gain_{nepers} \qquad \{A-11\}$$

Cable Attenuation db/100 ft to Nepers/Meter

$$\alpha_{nepers/meter} = 0.0037772065\,1737868\,\frac{neper\ feet}{db\ meter}\,\alpha_{db/100ft}$$

$$\{A-12\}$$

Cable Attenuation Nepers/Meter to db/100 ft

$$\alpha_{db/100feet} = 264.745916168\,\frac{db\ meter}{neper\ feet}\,x\,\alpha_{nepers/meter} \qquad \{A-13\}$$

⌘ ⌘ ⌘

APPENDIX B DERIVATION OF AN L-C MATCHING NETWORK

The input impedance of a resonant vertical antenna is resistive or nearly so. This resistance is composed of leakages and wire losses, etc., but mostly, it is the radiation resistance of the antenna. This is the quantity of greatest interest since it is a measure of the antenna's effectiveness as a radiator. Unfortunately, the radiation resistance of HF verticals is often smaller than desired and varies with the frequency of the input signal and with the height of the antenna.

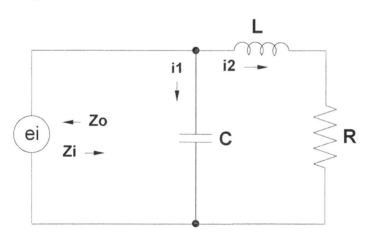

Figure B-1 Equivalent Circuit of Network & Antenna with Zo > R

Antenna modeling shows that the radiation resistance of a ground mounted quarter wavelength vertical antenna is no more than 36 ohms and will be less should the antenna be shorter than a quarter wave length at the operating frequency. This makes for a poor match for the 50 ohms of the usual transmission line used to feed it. The conventional L – C network works well over a limited frequency range in matching vertical antennas to transmission lines.

The following shows a derivation for the equations from which the needed values of L and C can be determined. Let fig. B-1 represent the situation.

Assume the normal case where the radiation resistance R is less than the impedance of the equivalent source which drives the antenna. There are two loop currents. The first flows through C:

$$e_i = i_1\left(\frac{-j}{\omega C}\right)$$

or

$$i_1 = -\frac{\omega C e_i}{j} = j\omega C e_i$$

and the second current flows through the radiation resistance R and the matching inductor L:

$$e_i = i_2(R + j\omega L)$$

or

$$i_2 = \frac{e_i}{R + j\omega L} = \frac{e_i(R - j\omega L)}{R^2 + \omega^2 L^2}$$

The impedance seen by the transmission line is then:

$$Z_i = \frac{e_i}{i_1 + i_2} = \frac{e_i}{j\omega C e_i + \dfrac{e_i(R - j\omega L)}{R^2 + \omega^2 L^2}}$$

$$= \frac{R^2 + \omega^2 L^2}{j\left(R^2 + \omega^2 L^2\right)\omega C + R - j\omega L}$$

$$= \frac{R^2 + \omega^2 L^2}{R + j\left(R^2 \omega C + \omega^3 L^2 C - \omega L\right)}$$

$$= \frac{\left(R^2 + \omega^2 L^2\right)\left[R - j\left(R^2 \omega C + \omega^3 L^2 C - \omega L\right)\right]}{R^2 + \left(R^2 \omega C + \omega^3 L^2 C - \omega L\right)^2}$$

Then carrying out the multiplication:

$$Z_i = \frac{R\left(R^2 + \omega^2 L^2\right)}{R^2 + \left(R^2 \omega C + \omega^3 L^2 C - \omega L\right)^2}$$
$$- \frac{j\left(R^2 + \omega^2 L^2\right)\left(R^2 \omega C + \omega^3 L^2 C - \omega L\right)}{R^2 + \left(R^2 \omega C + \omega^3 L^2 C - \omega L\right)^2} \qquad \{B-1\}$$

From this one equation, two equations can be constructed from which the unknowns L and C can be found. For the perfect match, the transmission line will not "see" any reactive terms when "looking" into the total load presented by the antenna and the L - C network taken together. Thus, all reactive terms on the right side of equation {B-1} must be zero, i.e.:

$$0 = \frac{-j\left(R^2 + \omega^2 L^2\right)\left(R^2 \omega C + \omega^3 L^2 C - \omega L\right)}{R^2 + \left(R^2 \omega C + \omega^3 L^2 C - \omega L\right)^2}$$

Yielding

$$0 = R^2 \omega C + \omega^3 L^2 C - \omega L \qquad \{B-2\}$$

or

$$C = \frac{\omega L}{R^2 \omega + \omega^3 L^2} = \frac{L}{R^2 + \omega^2 L^2} \qquad \{B-3\}$$

The second equation from {B-1} is constructed by forcing the real part of {B-1} to match the output impedance of the transmission line, call it Ro:

$$Ro = \frac{\left(R^2 + \omega^2 L^2\right)[R]}{R^2 + \left(R^2 \omega C + \omega^3 L^2 C - \omega L\right)^2} \qquad \{B-4\}$$

Now substitute the solution for C from {B-3} into {B-4} and then working through the algebra:

$$Ro = \frac{\left(R^2 + \omega^2 L^2\right)[R]}{R^2 + \left(R^2 \omega \dfrac{L}{R^2 + \omega^2 L^2} + \omega^3 L^2 \dfrac{L}{R^2 + \omega^2 L^2} - \omega L \dfrac{R^2 + \omega^2 L^2}{R^2 + \omega^2 L^2}\right)^2}$$

$$R_o = \frac{\left(R^2 + \omega^2 L^2\right)R}{R^2 + \left(R^2 \omega \dfrac{L}{R^2 + \omega^2 L^2} + \omega^3 L^2 \dfrac{L}{R^2 + \omega^2 L^2} - \omega L\right)^2}$$

$$R_o = \frac{\left(R^2 + \omega^2 L^2\right)R}{R^2 + \dfrac{1}{\left(R^2 + \omega^2 L^2\right)^2}\left[R^2 \omega L + \omega^3 L^3 - \omega L\left(R^2 + \omega^2 L^2\right)\right]^2}$$

$$R_o = \frac{\left(R^2 + \omega^2 L\right)^3 R}{R^2\left(R^2 + \omega^2 L^2\right)^2 + \left[R^2 \omega L + \omega^3 L^3 - \omega L\left(R^2 + \omega^2 L^2\right)\right]^2}$$

and

$$Ro = \frac{R\left(R^2 + \omega^2 L^2\right)^3}{R^2\left(R^2 + \omega^2 L^2\right)^2} \qquad \{B-5\}$$

$$R_o = \frac{\left(R^2 + \omega^2 L^2\right)}{R}$$

then

$$L^2 = \frac{RoR - R^2}{\omega^2}$$

$$L = \frac{\sqrt{R_oR - R^2}}{\omega} \qquad \{B-6\}$$

Now by using $\{B-6\}$ we can eliminate L from $\{B-3\}$:

$$C = \frac{L}{R^2 + \omega^2 L^2} = \frac{\dfrac{\sqrt{RoR - R^2}}{\omega}}{R^2 + \omega^2 \left(\dfrac{RoR - R^2}{\omega^2}\right)}$$

and get:

$$C = \frac{\sqrt{RoR - R^2}}{\omega RoR} \qquad \{B-7\}$$

Note that $\{B-6\}$ and $\{B-7\}$ were derived for the case where the load or antenna impedance is smaller than the transmission line impedance. Label these Set A expressions.

Should one wish to match a load resistance which is greater than the transmission line resistance, simply move the capacitor C to the other side of the series inductor shown in fig. B-1. Then this new network shown in fig. B-2 can then be used to find values for L and C to match these new impedances. When this is done, some will say that something unexpected occurs. Read on.

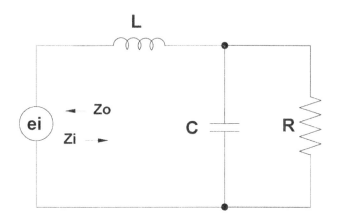

Figure B-2 Equivalent Circuit of Network & Antenna with Zo < R

Solving for L and C using this new schematic will yield expressions like those shown in {B-6} and {B-7}; label this second set as Set B. But in fact, it turns out that if the transmission line resistance is smaller than the load resistance, then by merely switching the labels of the two resistances in the Set A expressions will produce expressions exactly equal to the Set B equations!

A little thought about these so-called two different circuits, however, soon leads to the correct conclusion: the two different circuit diagrams are in fact the same, mathematically, when the labels of the two resistances are exchanged.

⌘ ⌘ ⌘

APPENDIX C COPY OF EXAMPLE 7-2D

	A	B	C	D	E	F	G	H	I
1		**Example 7-2d**							
2									
3		Find L & C for Match		Check Cal. Results					
4									
5		R	0.5	denomZi=	2.500E-01				
6		Ro	50	Re(Zi)num=	1.250E+01				
7		f=	3.855E+06	Im(Zi)num=	4.441E-14				
8		omega=	2.422E+07						
9				Re(Zi)=	5.000E+01				
10		L=	2.054E-07	Im(Zi)=	1.776E-13				
11		C=	8.216E-09						
12									
13		**Check Bandwidth of Match Network**							
14		freq	omega	denom	Re(num)	Im(num)	Re(Zi)	Im(Zi)	
15		3.845E+06	2.416E+07	2.506E-01	1.244E+01	6.331E-01	4.961E+01	2.526E+00	
16		3.847E+06	2.417E+07	2.504E-01	1.245E+01	5.074E-01	4.971E+01	2.026E+00	
17		3.849E+06	2.418E+07	2.502E-01	1.246E+01	3.812E-01	4.980E+01	1.523E+00	
18		3.851E+06	2.420E+07	2.501E-01	1.247E+01	2.546E-01	4.988E+01	1.018E+00	
19		3.853E+06	2.421E+07	2.500E-01	1.249E+01	1.275E-01	4.994E+01	5.101E-01	
20		3.855E+06	2.422E+07	2.500E-01	1.250E+01	4.441E-14	5.000E+01	1.776E-13	
21		3.857E+06	2.423E+07	2.500E-01	1.251E+01	-1.280E-01	5.005E+01	-5.119E-01	
22		3.859E+06	2.425E+07	2.501E-01	1.253E+01	-2.564E-01	5.008E+01	-1.025E+00	
23		3.861E+06	2.426E+07	2.502E-01	1.254E+01	-3.854E-01	5.011E+01	-1.540E+00	
24		3.863E+06	2.427E+07	2.504E-01	1.255E+01	-5.147E-01	5.012E+01	-2.056E+00	
25		3.865E+06	2.428E+07	2.507E-01	1.256E+01	-6.446E-01	5.013E+01	-2.572E+00	
26		3.867E+06	2.430E+07	2.509E-01	1.258E+01	-7.749E-01	5.012E+01	-3.088E+00	
27									
28		**Check Effect of Inductor Tolerance**							
29		f =	3.855E+06						
30		Omega =	2.422E+07						
31		C =	8.216E-09						
32	%	L	denom	Re(num)	Im(num)	Re(Zi)	Im(Zi)		
33	90	1.849E-07	4.421E-01	1.015E+01	8.896E+00	2.296E+01	2.012E+01		
34	93	1.900E-07	3.642E-01	1.071E+01	7.241E+00	2.942E+01	1.988E+01		
35	95	1.951E-07	3.036E-01	1.129E+01	5.228E+00	3.720E+01	1.722E+01		
36	98	2.003E-07	2.641E-01	1.189E+01	2.825E+00	4.501E+01	1.070E+01		
37	100	2.054E-07	2.500E-01	1.250E+01	-2.220E-14	5.000E+01	-8.882E-14		
38	103	2.105E-07	2.656E-01	1.313E+01	-3.281E+00	4.942E+01	-1.235E+01		
39	105	2.157E-07	3.156E-01	1.377E+01	-7.052E+00	4.363E+01	-2.235E+01		
40	108	2.208E-07	4.047E-01	1.443E+01	-1.135E+01	3.564E+01	-2.804E+01		
41	110	2.259E-07	5.381E-01	1.510E+01	-1.621E+01	2.806E+01	-3.012E+01		
42									
43		**Check Effect of Capacitor Tolerance**							
44		L =	2.054E-07						
45	%	C	denom	Re(num)	Im(num)	Re(Zi)	Im(Zi)		
46	90	7.394E-09	4.975E-01	1.250E+01	1.403E-01	2.513E+01	2.821E-01		
47	93	7.600E-09	3.897E-01	1.250E+01	1.059E-01	3.208E+01	2.718E-01		
48	95	7.805E-09	3.125E-01	1.250E+01	7.108E-02	4.000E+01	2.274E-01		
49	98	8.010E-09	2.659E-01	1.250E+01	3.578E-02	4.700E+01	1.345E-01		
50	100	8.216E-09	2.500E-01	1.250E+01	-8.314E-09	5.000E+01	-3.326E-08		

	A	B	C	D	E	F	G	H	I
51	103	8.421E-09	2.647E-01	1.250E+01	-3.627E-02	4.722E+01	-1.370E-01		
52	105	8.626E-09	3.100E-01	1.250E+01	-7.305E-02	4.032E+01	-2.356E-01		
53	108	8.832E-09	3.860E-01	1.250E+01	-1.104E-01	3.239E+01	-2.859E-01		
54	110	9.037E-09	4.926E-01	1.250E+01	-1.482E-01	2.538E+01	-3.009E-01		
55									
56									
57		**Check Bandwidth of Match Network**							
58		L=	2.054E-07						
59		C=	8.216E-09						
60		freq	VSWR	Pwr in Ant					
61		3.845E+06	1.053	9.878					
62		3.847E+06	1.042	9.881					
63		3.849E+06	1.031	9.883					
64		3.851E+06	1.021	9.884					
65		3.853E+06	1.010	9.885					
66		3.855E+06	1.000	9.885					
67		3.857E+06	1.010	9.885					
68		3.859E+06	1.021	9.884					
69		3.861E+06	1.031	9.883					
70		3.863E+06	1.042	9.881					
71		3.865E+06	1.053	9.878					
72		3.867E+06	1.064	9.876					
73									
74		**Check Effect of Inductor Tolerance**							
75		f=	3.855E+06						
76		C=	8.216E-09						
77		%	L	VSWR	Pwr in Ant				
78		90	1.849E-07	2.606	7.924				
79		93	1.900E-07	2.075	8.677				
80		95	1.951E-07	1.636	9.309				
81		98	2.003E-07	1.282	9.734				
82		100	2.054E-07	1.000	9.885				
83		103	2.105E-07	1.282	9.734				
84		105	2.157E-07	1.636	9.309				
85		108	2.208E-07	2.075	8.677				
86		110	2.259E-07	2.606	7.924				
87									
88		**Check Effect of Capacitor Tolerance**							
89		f=	3.855E+06						
90		L=	2.054E-07						
91		%	C	VSWR	Pwr in Ant				
92		90	7.394E-09	1.990	8.801				
93		93	7.600E-09	1.559	9.413				
94		95	7.805E-09	1.250	9.763				
95		98	8.010E-09	1.064	9.875				
96		100	8.216E-09	1.000	9.885				
97		103	8.421E-09	1.059	9.877				
98		105	8.626E-09	1.240	9.771				
99		108	8.832E-09	1.544	9.433				
100		110	9.037E-09	1.970	8.830				

	A	B	C	D	E	F	G	H	I
101									
102			**Table 7-2. Summary of**						
103			**Matching Network Performance**						
104									
105									
106									
107									
108									
109									
110									
111									
112									
113									
114									
115									
116									
117									
118									
119									
120									
121									
122									
123									
124									
125									
126									
127									
128									
129									
130									
131									
132									
133									
134									
135									
136									
137									
138									
139									
140									
141									
142									
143									
144									
145									
146									
147									
148									
149									
150									

	J	K	L	M	N	O	P
1							
2							
3							
4							
5							
6							
7							
8							
9							
10							
11							
12							
13							
14	Zi		VSWRr	Pwr in Ant			
15	49.6149518651036+2.52568560144208i		1.053	9.878			
16	49.7121897581368+2.0260876344891i		1.042	9.881			
17	49.7994312361092+1.523419333398597i		1.031	9.883			
18	49.8765526584672+1.01797845826529i		1.021	9.884			
19	49.9434421579801+0.510068878902844i		1.010	9.885			
20	50+1.77635683940025E-013i		1.000	9.885			
21	50.0461389073388-0.511913847018561i		1.010	9.885			
22	50.0817843488631-1.02535403071393i		1.021	9.884			
23	50.1068747900954-1.53999826649973i		1.031	9.883			
24	50.121361904319-2.05552128037192i		1.042	9.881			
25	50.1252107428988-2.57159548668001i		1.053	9.878			
26	50.1183998637654-3.08789167700494i		1.064	9.876			
27							
28							
29							
30							
31							
32	Zi						
33	22.9557871539415+20.1226741676446i		2.606	7.924			
34	29.4151171640885+19.8818954621885i		2.075	8.677			
35	37.2016889566786+17.221327918523i		1.636	9.309			
36	45.014376466484+10.6961107499519i		1.282	9.734			
37	50-8.88178419700125E-014i		1.000	9.885			
38	49.4190176734762-12.3512163556301i		1.282	9.734			
39	43.6290656835583-22.3454891145758i		1.636	9.309			
40	35.6428409126346-28.0410829783957i		2.075	8.677			
41	28.0567869367599-30.121526419338i		2.606	7.924			
42							
43							
44							
45	Zi						
46	25.1256295848789+0.282072438436522i		1.990	8.801			
47	32.0773724908036+0.271816996629945i		1.559	9.413			
48	40.0008021342312+0.227449907024016i		1.250	9.763			
49	47.0043555428348+0.134527803558119i		1.064	9.875			
50	50.0000006115506-3.3257843735684E-008i		1.000	9.885			

	J	K	L	M	N	O	P
51	47.2241071526898-0.137017502525556i		1.059	9.877			
52	40.3201403824199-0.235620457089232i		1.240	9.771			
53	32.38590884129-0.285915993958217i		1.544	9.433			
54	25.3781327439476-0.300906179491852i		1.970	8.830			
55							
56							
57							
58							
59							
60							
61							
62							
63							
64							
65							
66							
67							
68							
69							
70							
71							
72							
73							
74							
75							
76							
77							
78							
79							
80							
81							
82							
83							
84							
85							
86							
87							
88							
89							
90							
91							
92							
93							
94							
95							
96							
97							
98							
99							
100							

REFERENCES

Cheng, David K. 1989. *Field and Wave Electromagnetics, 2nd Edition*. Reading, MA: Addison-Wesley

Chipman, Robert A. 1968. *Transmission Lines*. New York: McGraw-Hill.

Johnson, Walter C. 1950. *Transmission Lines and Networks*. New York: McGraw-Hill.

King, Ronald W. P. 1965. *Transmission Line Theory*. New York: Dover Publications.

Kraus, John D. 1953. *Electromagnetics*. New York: McGraw-Hill.

Maxwell, James Clerk. 1873. *A Treatise on Electricity and Magnetism, Vol. 1 and 2*. Oxford: Macmillan.

Smith, Phillip H., "Transmission Line Calculator," *Electronics*, January 1939.

Smith, Phillip H., "An Improved Transmission-Line Calculator," *Electronics*, January 1944.

Straw, R. D., ed. *The ARRL Antenna Book, 20th Ed*. Newington, CT: American Radio Relay League, 2005.

⌘ ⌘ ⌘

INDEX

Index

Made in the USA
Coppell, TX
27 December 2023

26915284R00085